T0275654

Hybrid
RETROSYNTHESIS

Hybrid
RETROSYNTHESIS
Organic Synthesis using Reaxys and SciFinder

JOHN D'ANGELO

MICHAEL B. SMITH

ELSEVIER

AMSTERDAM • BOSTON • HEIDELBERG • LONDON
NEW YORK • OXFORD • PARIS • SAN DIEGO
SAN FRANCISCO • SINGAPORE • SYDNEY • TOKYO

Elsevier
Radarweg 29, PO Box 211, 1000 AE Amsterdam, Netherlands
The Boulevard, Langford Lane, Kidlington, Oxford OX5 1GB, UK
225 Wyman Street, Waltham, MA 02451, USA

Notices
Knowledge and best practice in this field are constantly changing. As new research and
experience broaden our understanding, changes in research methods, professional practices,
or medical treatment may become necessary.

Practitioners and researchers must always rely on their own experience and knowledge in
evaluating and using any information, methods, compounds, or experiments described herein.
In using such information or methods they should be mindful of their own safety and the safety
of others, including parties for whom they have a professional responsibility.

To the fullest extent of the law, neither the Publisher nor the authors, contributors, or editors,
assume any liability for any injury and/or damage to persons or property as a matter of products
liability, negligence or otherwise, or from any use or operation of any methods, products,
instructions, or ideas contained in the material herein.

ISBN: 978-0-12-411498-2

British Library Cataloguing-in-Publication Data
A catalogue record for this book is available from the British Library.

Library of Congress Cataloging-in-Publication Data
A catalog record for this book is available from the Library of Congress.

For Information on all Elsevier Publishing visit
our website at http://store.elsevier.com/

Typeset by MPS Limited, Chennai, India
www.adi-mps.com

Printed and bound in the USA

CONTENTS

PREFACE

Arguably, organic synthesis is one of the more difficult concepts for students to master in an organic chemistry course. Indeed, an advanced undergraduate course or a graduate-level course is often devoted to this single area of organic chemistry. One problem appears to be difficulty in learning and applying many different reactions. Another problem is learning and applying the principles of retrosynthetic analysis, but this issue is often exacerbated by the first problem. There are certainly many books that discuss and teach organic synthesis, and many books using slightly different approaches to teach undergraduate organic chemistry. Why is another book necessary?

Computer searching tools are ubiquitous, and several search engines are currently available to an organic chemist. Two of them, Reaxys and SciFinder, provide the capability of finding a given compound, a specific reaction, or examples of a specific transformation, and total syntheses may be found that have been reported in the literature. Such computer searches are less useful for assembling a total synthesis, although several computer programs have been developed to accomplish this task: LHASA,[1] CHIRON,[2] ArChem,[3] SYNGEN, and MARSEIL/SOS, among others. Most of these programs are rather sophisticated and expensive, however, and demand a level of chemical knowledge not yet attained after the typical undergraduate mainstream organic course sequence.

This book presents a relatively simple approach to solving synthesis problems, and a small library of basic reactions in Chapter 8, along with the computer searching capabilities of Reaxys and SciFinder. The idea is to use retrosynthesis to find a disconnect product for which a computer search will generate the remainder of the retrosynthesis. We term this approach a *hybrid retrosynthesis approach* since it uses the basics of retrosynthetic analysis to obtain a working intermediate compound that is subsequently searched with the computer. This book opens with an introduction to the retrosynthetic approach and a discussion of approaches for determining important bonds for disconnection. It ends with a brief discussion of more complex molecules and how this simple approach can be used with those challenging problems.

---

[1] See Lee, T.V. *Chemometrics and Intelligent Laboratory Systems* **1987**, *2*, pp. 259–272.
[2] See http://osiris.corg.umontreal.ca/chiron.shtml
[3] See http://www.archemcalc.com

Finally, for selected chemical transformations, a sample experimental procedure and the source from which it was gleaned is provided as the final section of this book. This section is not intended to supplant literature searching, but rather to provide a practical method for the reader to complete synthetic problems. In all cases, we strongly urge that the reader consult the primary literature found herein for the actual experimental protocol and also to search for additional examples that best meet his or her needs.

The authors thank Dr. Adrian Shell, formerly of Elsevier, who proposed an early version of this book, and who was responsible for the early research that led to this book. We further thank Ms. Jill Cetel and Ms. Katey Bircher, who have worked with us to develop the manuscript in its present form, and who have made production of this book possible.

Where there are errors, we take full responsibility. Please contact either of us if there are questions or errors. We hope that this book will be of value to both undergraduate organic chemistry students and first year graduate students taking an organic synthesis course.

John G. D'Angelo   **Michael B. Smith**
Alfred University   *University of Connecticut*

May 2015

COMMON ABBREVIATIONS

AIBN	*azo-bis*-isobutyronitrile	
All	Allyl	
Am	Amyl	$-CH_2(CH_2)_3CH_3$
aq	Aqueous	
	9-Borabicyclo[3.3.1]nonyl	
9-BBN	9-Borabicyclo[3.3.1]nonane	
BINAP	*2R,3S*-2,2′-*bis*-(diphenylphosphino)-1,1′-binapthyl	
Bn	Benzyl	$-CH_2Ph$
Boc	*t*-Butoxycarbonyl	
Bu	*n*-Butyl	$-CH_2CH_2CH_2CH_3$
Bz	Benzoyl	
c–	Cyclo-	
cat.	Catalytic	
Cbz	Carbobenzyloxy	$-CO_2CH_2Ph$
CIP	Cahn-Ingold-Prelog	
COD	1,5-Cyclooctadienyl	
Cp	Cyclopentadienyl	
Cy (*c*-C_6H_{11})	Cyclohexyl	
°C	Temperature in Degrees Centigrade	
DABCO	1,4-Diazabicyclo[2.2.2]octane	
d	Day(s)	
DBN	1,5-Diazabicyclo[4.3.0]non-5-ene	
DBU	1,8-Diazabicyclo[5.4.0]undec-7-ene	
DCC	1,3-Dicyclohexylcarbodiimide	c-$C_6H_{11}$$-N = C = N$-$c$-$C_6H_{11}$
DCE	1,2-Dichloroethane	$ClCH_2CH_2Cl$
% de	% Diastereomeric excess	
DEA	Diethylamine	$HN(CH_2CH_3)_2$
DEAD	Diethylazodicarboxylate	$EtO_2C-N = NCO_2Et$
DET	Diethyl tartrate	
DIPT	Diisopropyl tartrate	
DMAP	4-Dimethylaminopyridine	
DME	Dimethoxyethane	$MeOCH_2CH_2OMe$

(Continued)

(Continued)

DMF	N,N'-Dimethylformamide	
DMS	Dimethyl sulfide	
DMSO	Dimethyl sulfoxide	
e^-	Electrolysis	
% ee	% Enantiomeric excess	
Et	Ethyl	$-CH_2CH_3$
EDA	Ethylenediamine	$H_2NCH_2CH_2NH_2$
EDTA	Ethylenediaminetetraacetic acid	
Equiv	Equivalent(s)	
GC	Gas chromatography	
gl	Glacial	
h	Hour (hours)	
$h\nu$	Irradiation with light	
HMPA	Hexamethylphosphoramide	$(Me_2N)_3P = O$
HMPT	Hexamethylphosphorus triamide	$(Me_2N)_3P$
^{1}H NMR	Proton Nuclear Magnetic Resonance Spectroscopy	
HPLC	High performance liquid chromatography	
i-Pr	Isopropyl	$-CH(Me)_2$
IR	Infrared spectroscopy	
LDA	Lithium diisopropylamide	$LiN(i\text{-}Pr)_2$
MCPBA	$meta$-Chloroperoxybenzoic acid	
Me	Methyl	$-CH_3$ or Me
Mes	Mesityl	$2,4,6\text{-tri-Me}-C_6H_2$
min	minutes	
Ms	Methanesulfonyl	$MeSO_2-$
MS	Molecular Sieves (3 Å or 4 Å)	
NBS	N-Bromosuccinimide	
NCS	N-Chlorosuccinimide	
NIS	N-Iodosuccinimide	
Ni(R)	Raney nickel	
NMO	N-Methylmorpholine N-oxide	
Ⓟ or  ●	Polymeric backbone	
PCC	Pyridinium chlorochromate	
PDC	Pyridinium dichromate	
PEG	Polyethylene glycol	
Ph	Phenyl	
PhH	Benzene	
PhMe	Toluene	
Phth	Phthaloyl	
PTSA	$para$-Toluenesulfonic acid	
Pr	n-Propyl	$-CH_2CH_2CH_3$
Py	Pyridine	

(Continued)

(Continued)

Quant	Quantitative yield	
rt	Room temperature	
sBu	*sec*-Butyl	$CH_3CH_2CH(CH_3)$
sBuLi	*sec*-Butyllithium	$CH_3CH_2CH(Li)CH_3$
s	seconds	
TBAF	Tetrabutylammonium fluoride	$n\text{-}Bu_4N^+ \ F^-$
TBDMS	*t*-Butyldimethylsilyl	$t\text{-}BuMe_2Si-$
TBHP (*t*-BuOOH)	*t*-Butylhydroperoxide	Me_3COOH
t-Bu	*tert*-Butyl	$-CMe_3$
TFA	Trifluoroacetic acid	CF_3COOH
TFAA	Trifluoroacetic anhydride	$(CF_3CO)_2O$
Tf (OTf)	Triflate	$-SO_2CF_3 = (-OSO_2CF_3)$
THF	Tetrahydrofuran	
THP	Tetrahydropyran	
TMEDA	Tetramethylethylenediamine	$Me_2NCH_2CH_2NMe_2$
TMS	Trimethylsilyl	$-Si(CH_3)_3$
Tol	Tolyl	$4\text{-}(Me)C_6H_4$
Tr	Trityl	$-CPh_3$
Ts(Tos)	Tosyl = *p*-Toluenesulfonyl	$4\text{-}(Me)C_6H_4SO_2$
UV	Ultraviolet spectroscopy	

SUMMARY OF REACTIONS IN CHAPTER 8

(*Continued*)

(Continued)

(Continued)

(*Continued*)

(Continued)

(*Continued*)

(Continued)

CHAPTER 1

Disconnections and Synthesis

Contents

1.A THE DISCONNECTION APPROACH

To many students, the study of organic chemistry is just a grueling test of memory where they are forced to memorize countless reactions, all apparently differentiated by the most innocuous details. When asked to use those reactions to prepare new molecules, which is known as synthesis, one must have the ability to remember and use those reactions for a specific ways. One important question that students ask is, why is it necessary to know so many reactions? In fact, it is not an unfair question. The incredible diversity of structures and functional groups that appear in synthetic targets make a detailed knowledge of reactions an absolute necessity. It is probably true that an understanding of so many different reactions can only be understood by years of study, accompanied by laboratory work.

The diversity of structures leads to many synthesis problems because different carbon—carbon bonds must be made, and the incorporation of many different functional groups leads to different problems. One target molecule may contain several groups that are identical, such as a molecule with six different alcohol units, or three different ketone units. Such diversity makes synthesis virtually impossible without a thorough understanding of many different reactions, and why certain ones are better choices under specific conditions. It is therefore important to understand several different ways to make or incorporate the same functional group. On the plus side, a working knowledge of synthesis and protocols used to synthesize molecules can help with remembering reactions and also understanding them. Students in their first organic chemistry course often struggle with correlating the correct reaction and reagent with the correct functional group and may not appreciate this observation. This book attempts to fix this failure to see the forest through the trees, at least for

Hybrid Retrosynthesis.
DOI: http://dx.doi.org/10.1016/B978-0-12-411498-2.00001-2

those who have completed the two-semester organic chemistry sequence. To that end, the discussions in this book will provide a working library of common reactions and protocols used to synthesize molecules.

The chemical structure of medicines and other important molecules are characterized by the presence of many carbon atoms and often several functional groups. If such a molecule is not readily available from a commercial source, or a preparation is not found in the literature, a synthesis must be devised for that molecule. Normally, any synthesis requires choosing a commercially available molecule of fewer carbons as a starting material, and building the molecule, step-by-chemical step. Building a molecule in this manner is known as chemical *synthesis*, and it requires making carbon—carbon bonds to convert a smaller molecule into a larger and more complex one. Arguably, the fastest way to find out how well one understands reactions in organic chemistry is to attempt a synthesis that requires many different reactions. Planning a synthesis therefore instantly brings to light those reactions that are known and those that are not.

Traditionally, books and classroom studies provide the theoretical knowledge of organic reactions necessary to be a practicing organic chemist. Such knowledge is honed and refined in the laboratory and in the library to produce a true synthetic chemist, using knowledge of reactions, reagents, and theory to prepare organic compounds. If a planned reaction does not work, an alternative must be found, and/or research must be done to determine why the reaction did not work. All such studies require a good working understanding of organic chemistry in general, and organic reactions in particular. The power of computers and modern database searching is a great asset to the way in which modern chemists search for information in their quest to solve those synthetic problems. Indeed, the way in which chemists obtain and process knowledge has been modified and transformed by technology. In principle, this knowledge is used to modify an experiment to make it work. One can speculate on the advantage of learning reactions by computer searches. Faced with a problem in a total synthesis or any given reaction, there is no question that using computer searches to screen the chemical literature for reactions, or to find reactions for a given transformation is remarkably powerful.

This book will provide an overview of organic synthesis, and describe how to use two computer databases, REAXYS and SciFinder Scholar, to assist syntheses and find reactions. It is assumed that the reader has had at least two semesters of an undergraduate organic chemistry course before using the techniques described here.

The total synthesis of complex organic molecules demands a thorough knowledge of reactions. There are two major categories of reaction types. Reactions in one class make carbon–carbon bonds, and are called *carbon–carbon bond-forming reactions*. Reactions in the second class change one functional group into another, and they are called *functional group exchange reactions*.

Nowadays, the relationship of two molecules in a planned synthesis, say the preparation of **4** from **1**, is commonly shown using a device known as a *transform*, defined by Nobel laureate E.J. Corey[1] as: "the exact reverse of a synthetic reaction to a target structure." The *target structure* is the final molecule one is attempting to prepare (in this case, alcohol **4**). To begin the process, note that **4** has one more carbon atom when compared to **1**. Therefore, a carbon–carbon bond must be formed as part of the synthesis. This fact also means that any analysis that will lead back to the starting material must mentally break a carbon–carbon bond in the target. Removal of fragments by mentally breaking bonds is known as a *disconnection*, and the overall process that leads from target back to starting material is known as *retrosynthesis*.

A comparison of **1** and **4** in Figure 1.1 indicates that a methyl group must be added to C2 during the course of the synthesis. In this retrosynthesis, alcohol **4** is "simplified" by mentally breaking a C–C bond that connects one methyl group to the hydroxyl (OH)-bearing carbon. This mental exercise is known as a *bond disconnection*. In all cases, simplification means that the disconnect products have fewer carbon atoms when compared to the target. This observation is usually determined by the fact that the starting material for a synthesis has fewer carbon atoms than the final target, and any retrosynthesis works from the target back to the starting material. Therefore, each disconnection should simplify the target if possible. Simplification can involve disconnection of a large piece or a small piece. For complex molecules, disconnection of a large fragment is almost always preferred, but for small molecules such as **4**, loss of even one carbon atom constitutes simplification.

Any synthesis takes an available molecule (called the *starting material*) and transforms it by a series of reactions into a molecule that is required for some purpose (the *target*). For a molecule of any complexity, the reactions employed must include both *carbon–carbon bond forming reactions* and *functional group transformations*. The number and nature of the reactions

---

[1] Corey, E.J.; Cheng, X. *The Logic of Chemical Synthesis*, Wiley-Interscience, NY, **1989**.

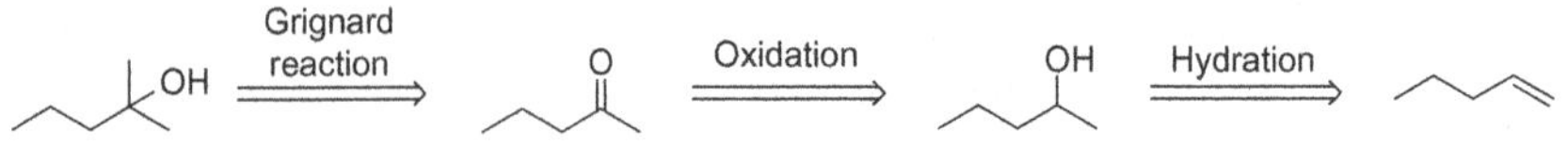

Figure 1.1 Retrosynthetic analysis of 2-methylpentan-2-ol (**4**).

are unknown by just looking at the target. Therefore, the purpose of the retrosynthetic analysis is to define both the number and nature of the requisite reactions for a given target.

This disconnected methyl carbon is attached to the OH-bearing carbon in **4**, and familiarity with chemical reactions suggests that this bond can be made by the reaction of a methyl reagent to a carbonyl group. In fact, this disconnection was chosen because it is known to the authors that an acyl addition reaction of a Grignard reagent to ketone **3** will give **4**. This knowledge arises from understanding reactions that change a ketone to an alcohol and others that change an alcohol to a ketone. Based on this disconnection, the retrosynthesis of **4** to **3** is shown as the synthetic precursor by a "backwards" arrow that is meant to show the synthetic relationship of **4** and **3**, as shown in Figure 1.1.

Once the disconnection of **4** generates disconnect product **3**, further analysis indicates that no C−C bond-forming reactions are required to pre-pare **3** from **1**. In other words, there are no additional carbon atoms in **3**, when compared to **1**. The remainder of the synthesis will involve only functional group exchange reactions. Functional group exchange reactions transform one functional group into another. Indeed, ketone **3** can be prepared from alcohol **2** by an oxidation and alcohol **2** can be prepared from alkene **1** by a hydration reaction. The synthetic relationships of these molecules are again shown by the "backwards" arrow, and each " ⇒ " is actually a transform that relates to the synthesis of **4** from **1**. In the retrosynthesis, the type of reaction planned to execute the transformation is often written above the ⇒ arrow, as in Figure 1.1. In this retrosynthesis, the plan is to convert **1** to **2** by a hydration reaction, **2** to **3** by an oxidation, and **3** to **4** via a Grignard reaction. This practice may help to overcome a temptation to write reagents over the ⇒ .

Working backward from **4** toward **1** in this manner is termed *retrosyn-thetic analysis* or *retrosynthesis*,[2] defined by Corey as "a problem-solving technique for transforming the structure of a synthetic target molecule to

---

[2] Reference 1, p. 6.

a sequence of progressively simple materials along a pathway which ultimately leads to a simple or commercially available starting material for chemical synthesis."[2]

The place to start a retrosynthetic analysis is with the compound one is trying to make, the target molecule. If a target must be prepared, several questions should be asked. *What is the starting material? What is the first chemical step? What reagents are used? How many chemical steps are required?* The answers to these questions are probably not obvious if more than one or two steps are required. The retrosynthesis protocol for analyzing the target first examines the target for ways to simplify the structure by a series of *mental bond-breaking steps* called *disconnections*. The term *disconnection* implies a thought experiment that breaks a bond within a molecule to generate simpler fragments, but no bonds are actually broken. The disconnections are based on an analysis of the structure, which must ultimately be based on knowledge of reactions that can construct those bonds and make those functional groups. In other words, when a bond is disconnected, that bond must eventually be made by a chemical reaction known to generate that bond. Therefore, if a bond is disconnected, *a chemical process must be available to make that bond*. In other words, *choosing a specific disconnection points toward a bond that must be made by a known chemical reaction*.

Note that the *first disconnection leads to a chemical reaction that is actually the last step of the synthesis*. In other words, the last step in a synthesis is always the one that generates the final target. For a complex target, this disconnection approach is repeated until a synthesis of that molecule is constructed, typically based on known reactions. Note, however, that even after performing all of the disconnections, the synthesis still must be constructed using known chemical reactions.

The retrosynthesis of **4** to **1** shown above obviously involves rather simple molecules, but it is intended to demonstrate the essentials of the disconnection approach. With practice, the retrosynthetic approach quickly becomes an integral part of how one thinks about molecules and their synthesis. In this example, one can ask why disconnection of a methyl group connected to the OH-bearing carbon to give **3** is an important bond to break? Earlier in the discussion, the answer to this question was based on a structural analysis of differences between **4** and **1**, which is completely valid. However, it will not always be so trivial. For example, what if **1** is not known to be reasonable starting material? It is therefore reasonable to ask if there is a general protocol that may be used to answer this question for other targets. A useful protocol is based on the idea that bonds connecting a heteroatom to a carbon atom are polarized, and identifying bond polarization can assist the choice of a bond disconnection.

The concept of nucleophilic and electrophilic atoms in ionic and polarized intermediates is well known. Polarized bond notation such as $C^{\delta+}-Br^{\delta-}$ and $C^{\delta-}-Li^{\delta+}$ is commonly used in describing the reactivity of such bonds. Seebach used structure **5** to describe a bond polarization model that can be correlated with chemical reactivity.[3] The sites marked *d* in **5** represent *donor* sites or *nucleophilic* atoms. The sites marked *a* are *acceptor* sites and correspond to *electrophilic* atoms. Be warned, however, that these general rules of donor and acceptor sites may require modification when there is more than one site of reactivity, as in the reaction of an epoxide with a nucleophile. Bond polarization induced by the heteroatom extends down the carbon chain, due to the usual inductive effects that are a combination of through-space and through-bond effects.[4] The electrophilic carbon adjacent to X (C1 for example) is designated C^a (an acceptor atom) since proximity to the δ^- electronegative atom (X) induces the opposite polarity. Similarly, C2 is a donor atom (C^d), but less polarized than X (this carbon is further away from the electrons that induce the bond polarization), and C3 is a weak acceptor atom. As a practical matter, the effect is negligible beyond C4 and will be ignored in this discussion.

5

Based on an analysis of many reactions, one can make the *assumption* that most organic reactions involve highly polarized species or even formally ionic reagents. Using this assumption, the three bonds shown in

[3] Seebach, D. *Angew. Chem. Int. Ed.* **1979**, *18*, 239.
[4] (a) Baker, F.W.; Parish, R.C.; Stock, L.M. *J. Am. Chem. Soc.* **1967**, *89*, 5677; (b) Golden, R.; Stock, L.M. *Ibid* **1966**, *88*, 5928; (c) Holtz, H.D.; Stock, L.M. *Ibid* **1964**, *86*, 5188; (d) Branch, G.E.K.; Calvin, M. *The Theory of Organic Chemistry*, Prentice Hall: NY, **1941**, Chapter 6; (e) Ehrenson, S. *Progr. Phys. Org. Chem.* **1964**, *2*, 195; (f) Roberts, J.D.; Carboni, C.A. *J. Am. Chem. Soc.* **1955**, *77*, 5554; (g) Clark, J.; Perrin, D.D. *Quart. Rev. Chem. Soc.* **1964**, *18*, 295.

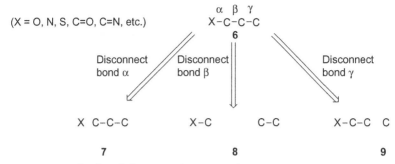

Figure 1.2 An α, β, γ-bond disconnection protocol.

Figure 1.2 become very important. Those three important bonds are the C−X bond (called the α bond), the adjacent one (the β bond), and the γ bond, as marked in **6**. If the α bond is disconnected, the synthetic step to form that bond requires direct attack of an X group (where X is a hetero-atom) on a suitably functionalized carbon (see **7** in Figure 1.2). Addition of an organometallic methyl group to a carbonyl fits this disconnection where X is C=O → CH$_3$−C−OH. If the β bond is disconnected, **8** is generated, along with C−X. In this case, the natural bond polarization of this bond suggests a reaction in which a nucleophilic carbon species attacks the electropositive carbon of the C−X unit, forming a new bond to the carbon that has the X group attached. Finally, disconnection of γ bond leads to **9** along with a carbon fragment. Conjugate addition of an organocuprate to an α,β-unsaturated ketone fits this type of disconnec-tion (where X in **9** is C=O, as in O=C−C=C reacting with a nucleophile C* to give **6**, O=C−C−C−C*).

Any planned retrosynthesis begins with a mental disconnection of a single bond, and any bond that is mentally broken must correlate with a reaction that can make that bond in a synthesis. The α, β, γ-bond disconnection protocol in Figure 1.2 is derived from an analysis of real reactions, so it is reasonable to assume that bonds α or β to the functional group (the OH unit of **4**) are good candidates for the first disconnection. Using this assumption, there are three bonds β to the carbon bearing the OH group, where the O−H bond is the α bond. Disconnection of one of the β bonds (C−CH$_3$ in Figure 1.3) leads to two disconnect fragments, **10** and **11**. There are two identical C−CH$_3$ bonds that lead to the same dis-connection. Disconnection of the remaining β bond (EtH$_2$C−CMe$_2$OH) generates disconnect fragments **12** and **13** in Figure 1.3.

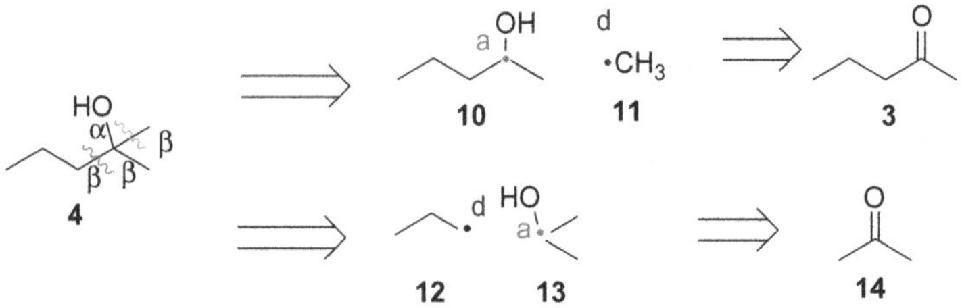

Figure 1.3 Disconnection of one β bond in 2-methylpentan-2-ol.

Fragments **10** and **11** as well as **12** and **13** are not real molecules. There must be a protocol that will correlate each disconnect fragment with a real molecule that will allow identification of a real organic reaction. Only then can a disconnection be classified as useful or unproductive. For all practical purposes, this statement means that the disconnection must be correlated with a known reaction with real molecules. In other words, the only way to determine if either disconnection is useful for a synthesis is to make a correlation with a real reaction.

Fragments **10** and **11** are termed *disconnect products*, and are not real molecules. Choosing the C−CH₃ bond disconnection in **4** is based on the fact that the bond is connected to the functional group-bearing carbon (see Figure 1.2). This disconnection leads to a methyl fragment (**11**) and an "alcohol" fragment (**10**).

When the disconnect products obtained by the first disconnection (of a C−CH₃ bond) are analyzed using the Seebach protocol, either **10** or **11** could be categorized as a donor (*d*) or an acceptor (*a*). Using the natural bond polarity associated with polarizing atoms, the choice is usually straightforward. A O−C fragment, for example is likely to be $^{\delta-}O-C^{\delta+}=\,^dO-C^a$ because oxygen is more electronegative than carbon. The bond polarization shown is, therefore, the "natural" bond polarization for the fragment. Based on bond polarization, it is more likely that the carbon of a C−O fragment would be an acceptor rather than a donor. Therefore, **10** is assigned with an acceptor carbon and **11** is assigned with a donor carbon, as shown in Figure 1.3. Using the same logic, the OH-bearing carbon atom in **13** is designated as an acceptor, and the carbon of propyl fragment **12** is designated as a donor.

To convert **10** and **11** into real molecules, a synthetic equivalent must be established for each disconnect product. Table 1.1 shows synthetic equivalents that correlate with donor and acceptor disconnect fragments. *Analysis of disconnect products using the list of synthetic equivalents shown in*

Table 1.1 Common Synthetic Equivalents for Disconnect Products

C^d or C^a	Synthetic equivalent
R-C·OR (a) and R-C·OH (a), with R	$\underset{R}{\overset{O}{C}}$·R or $-\overset{Cl}{\underset{OR}{C}}-H$
R,R C-CH$_2$ (a) =O and R,R C-CH$_2$ (a) with O-H ring	R,R C=C H, H with C=O
R,R C-C O,R (d) and R,R C-C-OH H, R R (d)	R,R C=C R, O$^-$ (enolate anion)
R,R C-C-OH R R R (a)	epoxide with R, R
R-C- R,R d	R-C-MgX R,R or R-C-Li R,R or $\left(R-C-\right)_2$CuLi R,R or R-PR$_3^+$ R,R
N-C- (d)	N≡C— or HN=CH— or H$_2$N·CH$_2$—
R-≡ (d)	R-≡-H or R,H C= H,H
R, C= (d) H, H	R-≡-H
R-C- R,R a	R-C-X R,R (X = Cl, Br, I, OTs, OMs, OTf, ...) (see abbreviations page)
R-C- R,R a	R-C-R O (for the Wittig reaction)

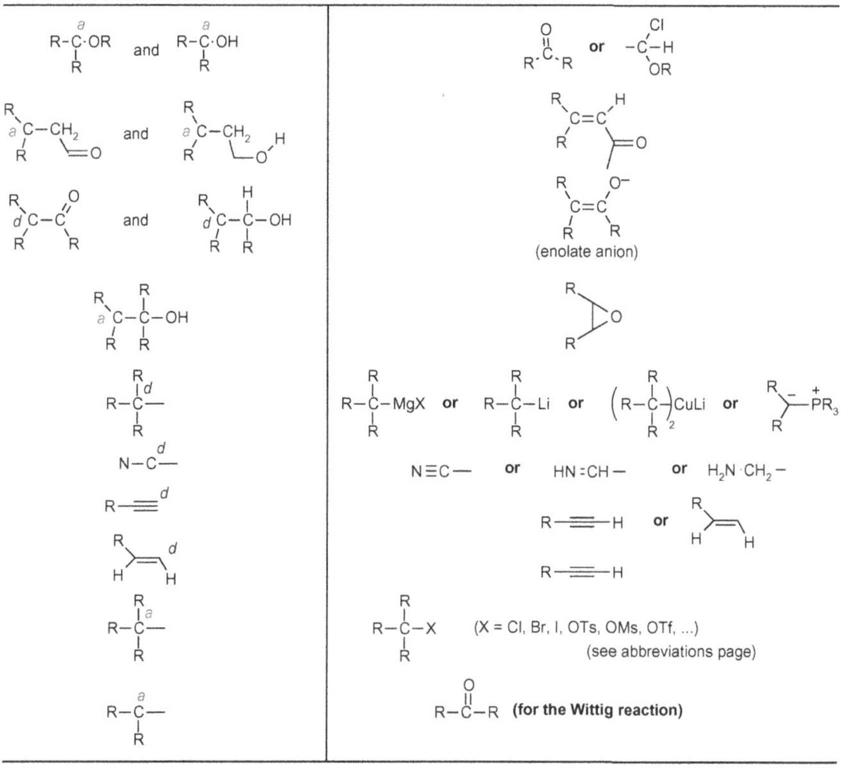

Table 1.1 *is essential for the retrosynthetic analyses generated in this discussion.* The synthetic equivalent for the O−C^{acceptor} fragment is an aldehyde or a ketone and that for O−C−C^{acceptor} is an epoxide.

The synthetic equivalents in Table 1.1 are based on polarization of a given bond using known functional groups and molecules that have the equivalent number of donor carbons or acceptor carbons. Those molecules with donor carbons include Grignard reagents or organolithium reagents, enolate anions, cyanide, and alkyne anions. Of these, only Grignard reagents and organolithium reagents do not have another functional group or a heteroatom, and this simple donor carbon has the synthetic equivalent C−MgX or C−Li. In addition to the simple cases represented by Grignard reagents and organolithium reagents, O=C−C^{donor} correlates with an enolate anion, N≡C^{donor} correlates with the cyanide anion, and C≡C^{donor} correlates with an alkyne anion.

Since the $C^{acceptor}$ is an electrophilic carbon, a reasonable synthetic equivalent may be an intermediate with a carbocation center, or a molecule with a δ^+ carbon. Several functional groups or substituents have this type of polarization. Two common classes of molecules polarized in this manner are alkyl halides and sulfonate esters, each with a $^{\delta-}X-C^{\delta+}$ unit. Molecules with this structural unit undergo S_N2 reactions with loss of X^-; the $C-X$ carbon can accept electrons from a donor atom. The polarized carbonyl group $(^{\delta-}O=C^{\delta+})$ also has the necessary bond polarization, and addition of a nucleophile to that carbon is consistent with a carbon accepting electrons from a donor nucleophile. The cases cited focus on a single atom, and bond polarization is based on direct attachment to a polarizing substituent. Bond polarization from a single polarizing substituent can extend to two adjacent carbon atoms rather than just one. An example is the three-membered ring of an epoxide, which has a $^{\delta+}C-^{\delta-}O-C^{\delta+}$ unit. The reaction of either of these polarized carbon atoms with a nucleophile leads to ring opening and formation of a $Nuc-C-C-O$ unit, which leads to the $^aC-C-O$ synthetic equivalent in Table 1.1.

Table 1.1 shows that a "simple" $C^{acceptor}$ with no heteroatom is best represented by an alkyl halide or a sulfonate ester. If an oxygen is connected directly to the acceptor carbon ($O-C^{acceptor}$ or $O=C^{acceptor}$), the synthetic equivalent is the carbonyl, but if the oxygen is on the adjacent carbon ($O-C-C^{acceptor}$), then the equivalent is the epoxide.

It is now possible to properly evaluate the disconnection in Figure 1.3. Disconnection of the indicated bond to the OH bearing carbon leads to **10** and the methyl fragment **11**. Bond polarization suggests that the $C-O$ unit in **10** should have an electrophilic carbon since oxygen is more electronegative. Therefore, that carbon is assigned as an acceptor carbon, and the synthetic equivalent for the ^aC-O unit **10** is a carbonyl. In this case, **10** correlates with a specific ketone, pentan-2-one, **3**. Although the methyl fragment **11** has no functional groups, it must correlate with C^d since fragment **10** correlates with C^a. Therefore, fragment **11** best correlates with a Grignard reagent, methylmagnesium bromide (CH_3MgBr). These reaction partners correlate with the well-known Grignard reaction, and knowledge of this reaction helped decide upon the disconnection of the important bond $HO-C\alpha-C\beta$. Further discussion for the correlation of known reactions with a retrosynthesis will be discussed in Chapter 2.

Before proceeding, it is useful to examine another disconnection to verify that using the concept of bond polarization for donor and acceptor

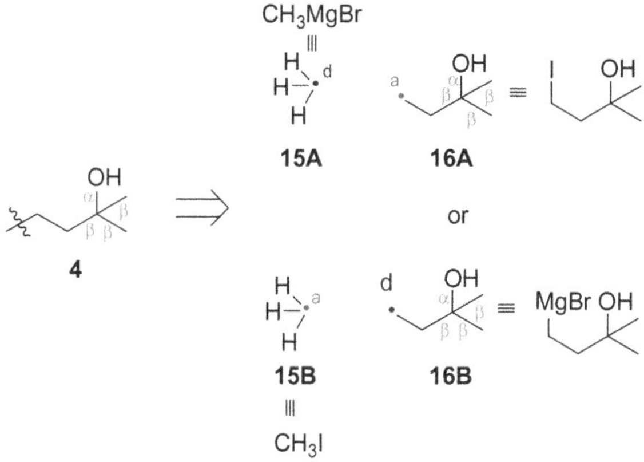

Figure 1.4 Disconnection of a distal bond in 2-methylpentan-2-ol, **4**.

sites as well as the idea of disconnection α or β to a functional group is valid. Returning to **4**, a different bond is disconnected in Figure 1.4, and *this bond is furthest away from the functional group* (the OH unit). Although structure **5** suggests that this terminal carbon should be a donor site, in fact, to an experienced synthetic chemist this bond is known to be a poor choice. Why is it a poor choice? This disconnection leads to two disconnect fragments, **15** and **16**. As shown, both can be assigned as either donor (**15A** and **16B**) or acceptor (**16A** and **15B**) sites since there is no bond polarization to guide the choice. Using Table 1.1, a reaction of a Grignard reagent or an organolithium reagent with an alkyl halide can be visualized from these fragments. Such an assignment suggests two reactions: (a) reaction of a iodoalcohol with methylmagnesium bromide, or (b) reaction of the OH-containing Grignard reagent with bromomethane or iodomethane. It is known that reactions of a Grignard reagent and an alkyl halide do not work well unless the halide is extremely reactive. An organocuprate can be used in a reaction with an alkyl halide rather than a Grignard reagent or an organolithium reagent, which will correct the reactivity problem, but the OH unit still poses a problem. Indeed, the acidic OH proton reacts as an acid with the potent base such as a Grignard reagent. It is clear from these comments, that *knowledge of all possible reactions is essential in order to make an informed decision relating to a disconnection.*

As noted above, a discussion of suitable reagents and reactions will be given in Chapter 2, but using our knowledge of organic reactions, it is clear that the disconnection in Figure 1.4 has problems whereas that in Figure 1.3 led to a straightforward synthesis based on known

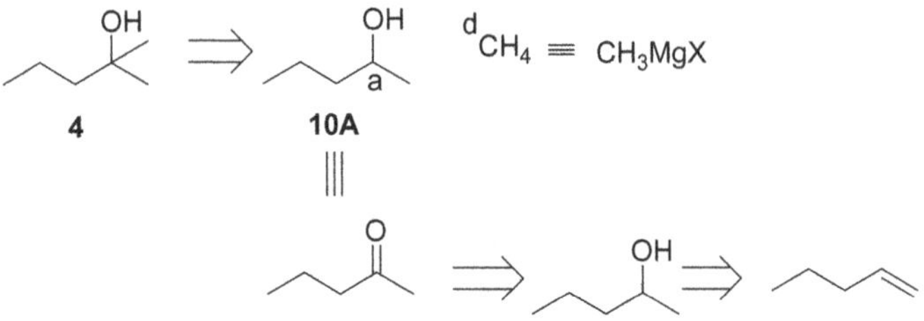

Figure 1.5 Progress in the retrosynthetic analysis of 2-methylpentan-2-ol (**4**).

chemical reactions. Such insight highlights the need for a deep understanding of organic reactions in choosing appropriate disconnections
and synthetic steps.

The retrosynthetic analysis that led to pentan-2-one (**3**) as the retrosynthetic precursor to **4** is not complete, because pent-1-ene (**1**) is the starting
material and only the last chemical step of the synthesis has been determined.
*The disconnection/simplification process must be repeated, but using **3** as the precursor.*
Note that **1**, **2**, and **3** all have five-carbon atoms, so no carbon–carbon
bond–forming reactions are required in any of the remaining steps that must
be considered. In other words, there is no need to disconnect any
carbon–carbon bonds. At this point, an understanding of functional group
interactions and their manipulation is essential to complete this retrosynthesis
plan, and then execute the actual synthesis. It is reasonable to ask if **1** is direct
a precursor to **3**. In other words, is there a reaction that converts **1** to **3** in a
single reaction? Based on undergraduate chemistry, the answer is probably
no! Therefore, to complete the synthesis one must ask how can one prepare
a ketone. One answer is the oxidation of an alcohol to a ketone. If this
choice is made, one must then ask how one makes an alcohol. Another
well-known reaction is the reaction of an alkene with aqueous acid (a hydration reaction). Unless one is familiar with these transformations, completion
of the synthesis will require a literature search to answer these questions
relating to functional group transformations.

As noted, the oxidation of an alcohol to a ketone is well known, so
choosing an alcohol as the precursor to **3** makes sense, *if one is familiar
with this transformation*. Using an oxidation reaction, the alcohol precursor
to **3** is pentan-2-ol (**2**), as shown in Figure 1.5. If **2** is the precursor to **3**,
it is reasonable to ask if **1** is a precursor to **2**. Alkenes are known to react
with an acid catalyst to form a more stable carbocation, so **1** will react to
form a secondary carbocation. Subsequent reaction with water and loss of

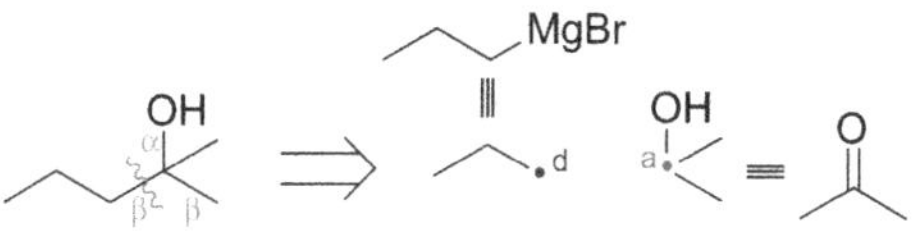

Figure 1.6 Alternative disconnection of **4**.

a proton from the intermediate oxonium ion gives **2**. Knowledge of this hydration reaction allows completion of the retrosynthesis shown in Figure 1.5. The interconversion of ketone−alcohol−alkene units involves modification of functional groups, it these reactions are classified as *functional group exchange reactions*.

It is useful to note that an alternative disconnection of **4** to give **12** and **13** would suggest a reaction between acetone (**14**) and propylmagnesium bromide, as shown in Figure 1.6. This disconnection is perfectly reasonable, but remember that the given starting material is pent-1-ene, so this approach cannot be used in this particular case.

1.B FUNCTIONAL GROUP EXCHANGE REACTIONS

The disconnection in Figure 1.5 involved a carbon−carbon bond, but an analysis of the complete syntheses makes it clear that many, if not most of the actual reactions manipulate functional groups. These are *functional group exchange reactions*, and it is a fact that manipulating functional groups is an essential component of synthesis, and sometimes accounts for all of the synthetic steps. Is there a way to organize functional group exchanges?

In most undergraduate organic chemistry textbooks, the functional group approach to introducing chemistry means that groups with related chemical properties are usually presented in different semesters. The chemical reactions of an alkene or a carbonyl are presented in different semesters, and in different chapters during a typical course. However, the π-bond of an alkene and that of a ketone or aldehyde both react with Brønsted−Lowry acids, forming a carbocation or an oxocarbenium ion, respectively. For all practical purposes, both can be classified as Brønsted−Lowry acid−base reactions. The bond polarization of the carbonyl also leads to the reaction of the carbonyl unit of aldehydes or ketones with nucleophiles via acyl addition, whereas the C=C unit of a simple alkene does not react with a nucleophile. This example is meant to illustrate that understanding the similarities and differences of functional groups is important for understanding the chemistry of those groups.

Perhaps the most common approach to teaching undergraduate organic chemistry is the so-called functional group approach. However, using the functional group approach tends to focus on the differences in functional groups, which can make the relationship of one functional group to another difficult to see. As just discussed, however, there are similarities in some functional groups that allow predictions of reactivity. Therefore, a mnemonic that gives a quick overview of the chemical relationship of functional groups is useful. Figure 1.7[5] provides a visual reminder of these relationships, which are often essential for completion of a total synthesis. Pertinent reactions for the transformations *a-x* are provided based on the numbered reactions in Chapter 8 and reactions for each transformation will be elaborated in Chapter 5.

In the example from Figure 1.5 that prepares **3** from **1**, the relationship between a ketone and an alcohol or an alcohol with an alkene is apparent from the table. Ketones can be prepared by oxidation of an alcohol ($C-C-OH \rightarrow C-C=O$) and an alcohol is prepared by reduction of a ketone ($C-C=O \rightarrow C-C-OH$). While alkenes can be prepared directly from alkyl halides via E1 reactions, this approach is not very efficient and there are often competing side reactions that lead to poor yields. Examination of Figure 1.7 shows that halides and alcohols are chemically related in that an alcohol can be converted to a halide and a halide to an alkene ($C-C-OH \rightarrow C-C-Br \rightarrow C=C$), but an alkene can be directly converted to an alcohol ($C=C \rightarrow C-C-OH$). *This hydration reaction of an alkene to an alcohol is not obvious in* Figure 1.7, although the scheme can be modified such that $C-C-X$ is used, where X=OH. Therefore, the $C=C \rightarrow C-C-X$ and $C-C-X \rightarrow C-C-OH$ are redundant and somewhat misleading. These groups were chosen to make the correlation scheme more general, but the schemes in Figure 1.7 are not all encompassing. Once again, the intent is to show that functional groups are related, which is a paradigm change from the way most undergraduate textbooks are organized. Figure 1.7 will be repeated in Chapter 5, where detailed searches will be presented to show how to find reaction.

Returning to the analysis of **3** in Figure 1.5, pent-1-ene can be converted to pentan-2-ol, which can be oxidized to pentan-2-one, which will complete the synthesis, as outlined in Figure 1.8. The retrosynthetic analysis of **4** suggested a Grignard reaction to convert **3** to **4**, but reagents are necessary for reactions that convert **1** to **2** and **2** to **3**. At this point those reagents are unknown, so **A** and **B** are used on the figure. A study of Figure 1.7, and a

[5] Smith, M.B. *J. Chem. Ed.* **1990**, 67, 848−856.

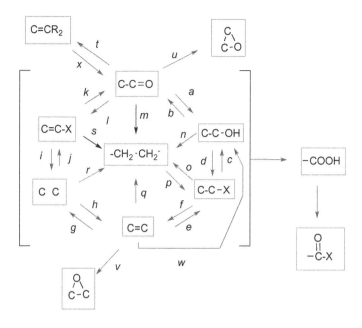

a. butan-2-one to butan-2-ol (06)
c. butan-2-ol to 2-bromobutane (39)
e. but-2-ene to 2-bromobutane (41)
g. but-2-ene to but-2-yne (see 41 and 19)
i. but-2-yne to 2-bromobut-2-ene (43)
k. 2-bromobut-2-ene to butan-2-one (45)
m. butan-2-one to butane (16)
o. 2-bromobutane to butane (15)
q. but-2-ene to butane (14)
s. 2-bromobut-2-ene to butane (14)
u. butan-2-one to 2-ethyl-2-methyloxirane (not listed)
w. but-2-ene to butan-2-ol (04)

b. butan-2-ol to 2-butan-2-one (09)
d. 2-bromobutane to butan-2-ol (see 41 and 04)
f. 2-bromobutane to but-2-ene (18)
h. but-2-yne to but-2-ene (17)
j. 2-bromobut-2-ene to but-2-yne (19)
l. butan-2-one to 2-bromo-but-2-ene (not listed)
n. butan-2-ol to butane (04)
p. butane to 2-bromobutane (40)
r. but-2-yne to butane (see 17 and 14)
t. butan-2-one to 2-methyl-but-1-ene (66)
v. but-2-ene to 2,3-dimethyloxirane (34)
x. 2-methylbut-1-ene to butan-2-one (10)

Figure 1.7 Functional group exchange reaction wheel. A visual reminder of the chemical relationships of functional groups. The numbers in parentheses, (04), refer to the pertinent reactions in Chapter 8. *Reprinted with permission from Smith, M.B. and the* Journal of Chemical Education, *Vol. 67, 1990, 848−856. Copyright 1990, Division of Chemical Education, Inc.*

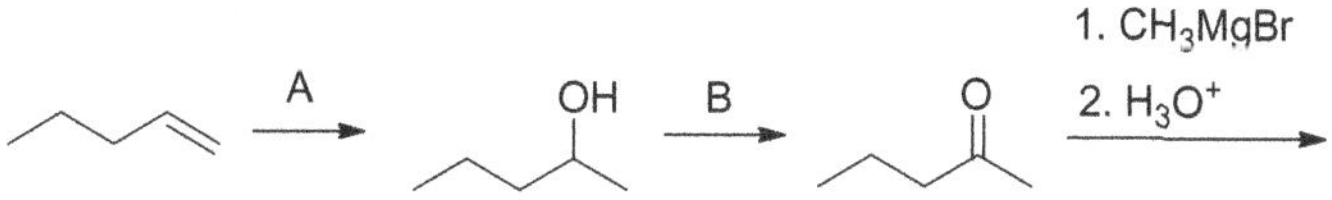

Figure 1.8 Synthesis of 2-methylpentan-2-ol from pent-1-ene.

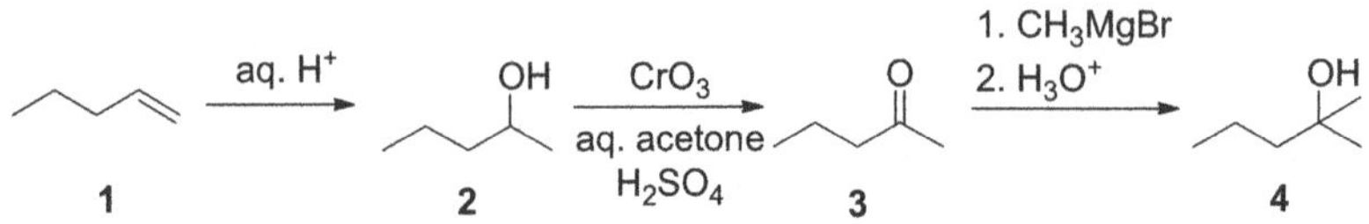

Figure 1.9 Synthesis of **4** from **1**.

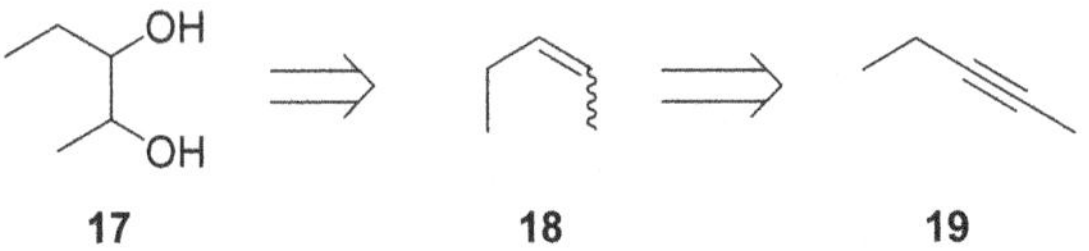

Figure 1.10 Retrosynthesis of pentane-2,3-diol from pent-2-yne.

review of most undergraduate organic chemistry courses, should allow one to identify at least one reagent to accomplish each transformation.

As a practical matter, and for the purposes of the reactions in Figure 1.8,[5] some help is needed to supply reagents. A list of reactions based on functional group transformations and carbon−carbon bond-forming reactions is presented in Chapter 8. Those reactions are organized by the functional group being made: alcohol, alkene, ketone, etc. Reactions 1−53 are functional group exchange reactions, whereas Reactions 54−82 are carbon−carbon bond-forming reactions. Using this focused library of reactions, reagents can be provided for the transformations in Figure 1.8. An understanding of how each transformation works (the mechanism) is essential before choosing a reagent, and a proper organic chemistry course is probably the best way to provide that knowledge. This information was briefly mentioned in this chapter to explain the disconnection in Figure 1.5. The transformation of **1** to **2** prepares an alcohol from an alkene. However, since the OH group in **2** resides on the secondary carbon, so it is likely that a carbocation mechanism is operative. Therefore, Reaction 04 (Chapter 8) is a proper choice. The conversion of **2** to **3** prepares a ketone from an alcohol, so Reaction 09 in Chapter 8 is a good choice. Any of those reagents will suffice, but Jones oxidation with CrO_3 is the simple choice. With these reagents in hand for the specific transformations, the synthesis can finally be completed, as shown in Figure 1.9.

There are syntheses that do not involve making carbon−carbon bonds, so there are no formal disconnections in those retrosynthetic pathways, only functional group exchange reactions. If pentane-2,3-diol (**17**) is the target, a retrosynthesis can be defined in which the starting material must be pent-2-yne (**19**), as shown in Figure 1.10. Analysis of

Figure 1.11 Synthesis of **17** from **19**.

these two structures reveals that the five-carbon atoms in the alkyne are maintained in the diol, so the synthesis will not require a carbon−carbon bond-forming reaction.

In Figure 1.10, the retrosynthetic precursor to the diol is pent-2-ene (**18**), as either the *E* or the *Z* isomer since the stereochemistry of the diol is undefined. The retrosynthetic precursor to the alkene is the starting material, pent-2-yne (**19**). The relationship of these groups is found in Figure 1.7, which shows that the alkene can be prepared from an alkyne (Reaction 17 in Chapter 8), and the diol can be prepared from an alkene (Reaction 31 in Chapter 8). Using catalytic hydrogenation to convert the alkyne to the alkene (Reaction 17 in Chapter 8) and potassium permanganate to prepare the diol (Reaction 31), a viable synthesis is shown in Figure 1.11. If the stereochemistry of the diol unit in **17** is important, then Lindlar's catalyst[6] should be used for the hydrogenation of **19**, giving (2*Z*)-pentene as the major product. Using Figure 1.7 in conjunction with the reactions in Chapter 8 proved to be a useful combination to tap reactions and reagents learned in an undergraduate organic chemistry course.

[6] Lindlar, H. *Helv. Chim. Acta* **1952**, *35*, 446.

CHAPTER 2

Making Carbon–Carbon Bonds

The disconnection of a molecule that has many C–C bonds will inevitably lead to simplification and the generation of molecules with fewer carbon atoms. The retrosynthetic analysis of a large molecule will involve several C–C bond disconnections to give smaller molecules, so that synthesis will require the use of carbon–carbon bond-forming reactions. In general, a retrosynthetic analysis is biased toward specific reactions. This observation raises the issue of how many carbon–carbon bond-forming reactions are taught in a typical undergraduate organic chemistry course? The answer to this question is usually 9–12 reactions. Twelve common carbon–carbon bond-forming reactions are listed below, along with the pertinent reactions that can be found in Chapter 8.

1. **S$_N$2 reactions of alkyl halides or alkyl sulfonate esters with cyanide ion or alkyne anions.**
   There are many S$_N$2 reactions that involve a variety of nucleophiles. Common S$_N$2 reactions listed in Chapter 8 include Reactions 67 and 81.

2. **S$_N$2 reactions of enolate anions (from ketones, aldehydes, esters) with alkyl halides or alkyl sulfonates.**
   Examples that form C–C bonds and use S$_N$2 reactions include Reactions 58 and 74.

3. **Substitution reactions of organometallics such as organocuprates with alkyl halides, alkyl sulfonates, and related compounds.**
   Reaction 62.

4. **Addition of carbanions and organometallics such as Grignard reagents and organolithium reagents to the carbonyl of aldehydes or ketones (acyl addition), and nitriles.**
   Reactions include 56, 57, 68, 75, 79, and 82.

5. **Acyl addition of enolate anions (from ketones, aldehydes, esters) to aldehydes or ketones; the aldol condensation.**
   Reaction 80 and see 78.

6. **Acyl substitution, including the Claisen condensation with esters.**
   Reactions include 54, 57, 71, 73, 75, and 76.

Hybrid Retrosynthesis.
DOI: http://dx.doi.org/10.1016/B978-0-12-411498-2.00002-4

7. **Conjugate addition of organocuprates to conjugated aldehydes, ketones, or esters.**
   Conjugate addition reactions include 56, 75, and 78.

8. **Friedel–Crafts alkylation and acylation with aromatic substrates.**
   Friedel-Crafts alkylation is found in Reaction 69 and Friedel-Crafts acylation is found in Reaction77.

9. **Wittig reactions with aldehydes or ketones to give alkenes.**
   The Wittig reaction is found in Reaction 66.

10. **Diels–Alder cyclization of dienes with alkenes.**
    The Diels-Alder reaction is found in Reaction 64.

11. **Sigmatropic rearrangements such as the Cope and Claisen rearrangement.**
    Sigmatropic rearrangements are represented by Reaction 72. The Cope rearrangement is found in Reaction 72 and the Claisen rearrangement is found in Reaction 60.

12. **Transition metal-catalyzed (i.e., Pd, Ru) C−C bond-forming reactions.**
    There are many examples of reactions of this type. Examples include the Heck reaction (63), ring closing metathesis (65), the Sonogashira reaction (67), and the Suzuki-Miyaura coupling (70).

There are many other reactions that generate carbon−carbon bonds, and they are found in the chemical literature relating to organic chemistry. Any modern graduate course in organic chemistry will certainly expand the list of possibilities. The intent of this short list and reactions 54−82 in Chapter 8 is to provide a basic library of reactions that most chemists will have after an undergraduate course.

In the retrosynthesis of **4** in Figure 1.1 in Chapter 1, the retrosynthesis was biased towards a specific starting material (pent-1-ene), so the only disconnections examined are those involving bonds connected to the carbon atoms of the starting material. If no starting material is designated for a given target, the disconnection approach is used to develop a retrosynthetic analysis. The target is disconnected until a molecule is identified that can be purchased commercially, or one that is readily available.

If there is no designated starting material, there is more latitude in terms of the disconnections because there is no requirement to correlate a specific bond in the target with one in a particular starting material. Disconnections are driven by bond polarization related to functional groups, as previously described, and by the specific knowledge of the chemist performing the synthesis. Since the disconnection of a carbon−carbon bond will be important equivalents for C^{donor} in Table 1.1 are Grignard reagents, organolithium reagents, and organocuprates. The equivalent for $C\equiv C^{donor}$ or $C=C^{donor}$ is an alkyne anion and that for NC^{donor} is cyanide. The equivalent for $O-C-C^{donor}$ is an enolate anion. For simple $C^{acceptor}$ the synthetic equivalent is an alkyl halide or a sulfonate ester. The equivalent for $O-C^{acceptor}$ is an aldehyde or a ketone and that for $O-C-C^{acceptor}$ is an epoxide. In most synthetic problems, the reactions that make those bonds take on added importance. Note that these equivalents correlate with the need to disconnect carbon–carbon bonds near the functional group.

A target is shown in Figure 2.1, compound **1**, and the bonds are labeled *a-h*. The π-bond of the carbonyl (C=O) is not labeled since it is a functional group, nor are the π-bonds of the benzene ring, which is also considered to be a functional group for the purposes of the analysis. *Assume that carbon-carbon π-bonds, including those in benzene rings are not disconnected.* This assumption is made because the few reactions discussed in a typical undergraduate organic chemistry course do not present such reactions. Indeed the benzene ring is used as an intact entity (a unique functional group).

Disconnection of carbon–carbon π-bonds in an alkene is possible, and one prominent example is the Wittig reaction (Chapter 8, Reaction 66). Since there are relatively few ways to directly form a C=C bond, such disconnections will not be used very often in this book. The labeled bonds are subdivided into two categories: *the α bond that is connected directly to the functional groups and/or heteroatom, and the β bond that is one removed from the functional group or heteroatom.* In **1A**, the α-bonds to the carbonyl (C=O is one functional group and the phenyl ring is a second functional group) are labeled. In **1B**, the β-bonds to those two functional groups are labeled. These bonds are identified in order to choose one for the first disconnection. In principle, any bond is suitable, but there are clues in the structure of the molecule that suggests disconnection of some bonds is more productive.

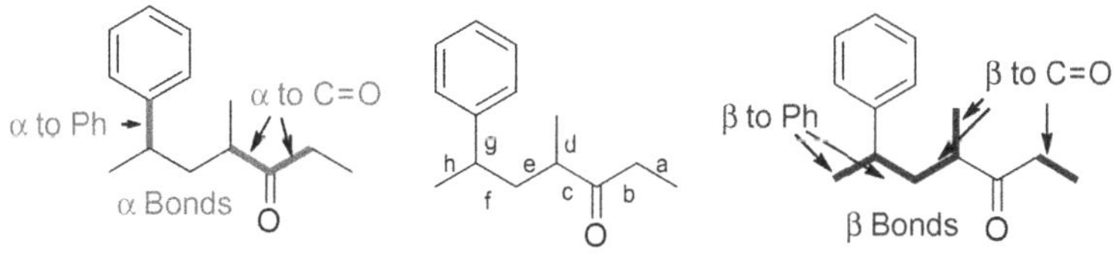

Figure 2.1 Key bonds in target **1**.

An important criterion for a disconnection is simplification of the target to the greatest extent possible. If either bond *b* or bond *d* in **1** is disconnected, the result is one very large fragment and one very small fragment. Disconnection of bond *d* in Figure 2.2, for example, gives **2** and a one-carbon fragment **3**, and disconnection of bond *b* leads to **4** and a two-carbon fragment **5**. Compare these fragments with disconnection of bond *c*, also proximal to the functional group, which gives **6** and **7**. Disconnection of bond *e* gives fragments **8** and **9**. In the latter two cases, the fragments are close to the same size and these disconnections provide significant simplification in that both **8** and **9** have fewer carbons but each fragment is approximately equal in size. This observation contrasts with fragments **4** and **5**, which have fewer carbon atoms than the targets, but there is a great disparity in the size of each fragment. In other words, disconnection of bond *d* or *b* removes small pieces whereas disconnection of bond *e* or *c* removes large pieces from the target.

A second important characteristic in a target is the presence of one or more stereogenic centers (see Chapter 6). Target **1** has two stereogenic centers. *Note that the racemic molecule is shown as the target.* The disconnection process begins with a focus on bonds that are proximal to the functional groups, which in this case is the carbonyl functional group or the benzene ring. This choice is clear by simply recognizing that there are many reactions involving a carbonyl that make a carbon−carbon bond, but few if any reactions involving benzene. *Note that one bond α to C=O (bond c) and two*

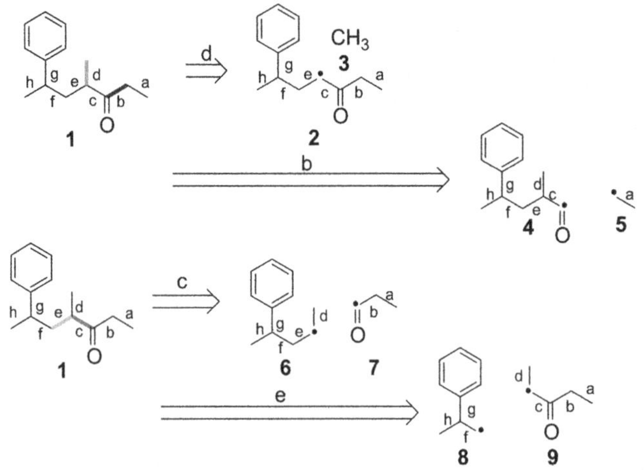

Figure 2.2 Two disconnections of target **1**.

bonds β to C=O (bonds e and d) are connected to the stereogenic center adjacent to the carbonyl. When possible, disconnect a bond that is connected to a stereogenic center. There is a good rationale for this statement. Formation of the stereogenic center during a reaction offers the potential to control the stereochemistry of that center. Disconnection of a bond that is not attached to a stereogenic center simply means that the stereogenic center must be made in another reaction. The stereogenic center in the fragment must be dealt with sooner or later. Sooner is better than later in a synthesis. Note that *the synthesis will generate racemic products and the focus is not on controlling absolute stereochemistry. Therefore, diastereoselectivity is an important issue in this analysis, but not enantioselectivity.* Bonds *f*, *g*, and *h* are also connected to a stereogenic center. Less simplification occurs by disconnecting these bonds so they are ignored until later in the synthesis.

Based on the preceding section, it appears that bonds *c* and *e* in target **1** are the best candidates for a disconnection. The disconnection of bond *e* in Figure 2.2 gives **8** and **9** and disconnection of bond *c* gives fragments **6** and **7**. As noted earlier, **6–9** are disconnect fragments and not real molecules. Real molecules must be obtained in order to ascertain if these are reasonable disconnections. Using the synthetic equivalents protocol from Chapter 1, each fragment is categorized as a donor or an acceptor. Fragment **7** is a O=C—C fragment, so it is logically an acceptor, O=C^a—C. This assignment correlates well with a carbonyl, specifically an aldehyde (propanal, **11**). This assignment also means that **6** must be the donor fragment, leading to the Grignard reagent derived from halide **10** (reaction 50 in Chapter 8). Assume that bromide **10** is not commercially available, and does not fit the criterion for a starting material, which means that another disconnection is required (Figure 2.3).

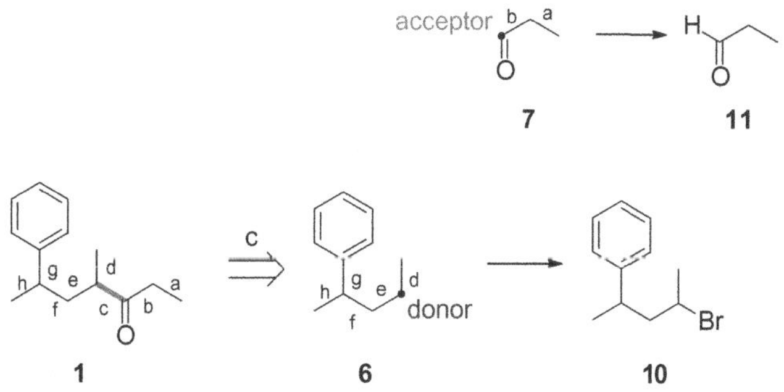

Figure 2.3 Disconnect product correlation for **1**.

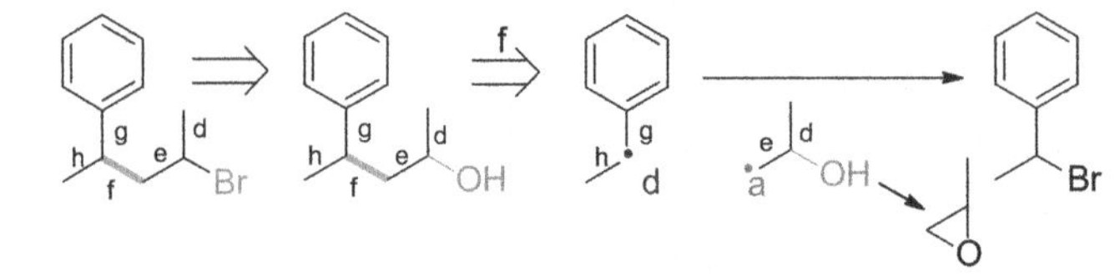

Figure 2.4 Refunctionalization and disconnection of bond *f*.

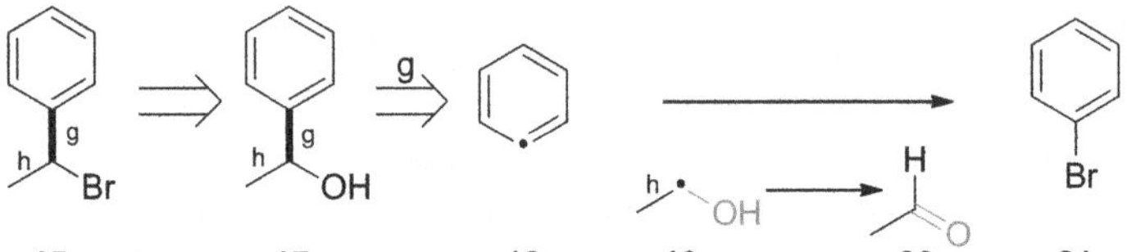

Figure 2.5 Refunctionalization and disconnection of bond g.

The halide **10** is not a good candidate for a disconnection because halogen substituents are usually treated as substituents rather than functional groups. However, changing the bromine group to an alcohol (see Figure 2.4) generates a new and viable target, **12**. Several possible disconnections are now possible based on the C—O bond polarization, and one is cleavage of bond *f* that leads to **13** and **14**. This disconnection was chosen because Table 1.1 contains the [acceptor]C—C—O fragment that correlates with an epoxide. This assignment correlates with acceptor fragment **14**, which has epoxide **16** as a synthetic equivalent. If **14** is the acceptor fragment, then **13** is the donor, which correlates with the Grignard reagent derived from benzylic halide **15**. Halide **15** may be a starting material if it is readily available. Many undergraduate, and some graduate textbooks offer synthetic problems where the starting material must be ≤five- or six-carbon atoms. This restriction is, of course, arbitrary and both the chemical literature and commercial chemical supply companies must be consulted to determine is a molecule is a suitable starting material.

As above, change the halide to an alcohol in the retrosynthesis (see Figure 1.7 in Chapter 1), making alcohol **17** the new target in Figure 2.5. If bond *g* is disconnected, the result is phenyl fragment **18** and the C—C—OH fragment, **19**. If this latter fragment is identified as the acceptor, it correlates with C—C[acceptor]—OH with a carbonyl as an equivalent, specifically acetaldehyde **20**. This assignment means that **18** is the donor and its equivalent is the Grignard reagent derived from bromobenzene **21**. Compound **21** is bromobenzene and is likely available as a starting material,

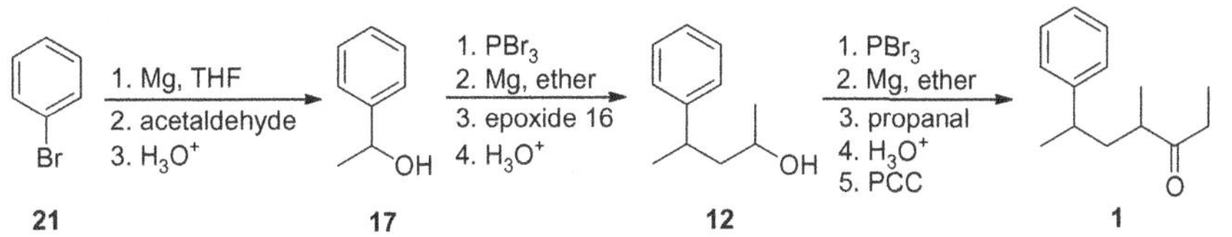

Figure 2.6 Synthesis of **1** from bromobenzene (**21**).

and for this example it is the starting material for the entire synthesis, based on the retrosynthetic scheme shown.

It is important to emphasize that *there is no "correct" disconnection for this molecule.* Choices are made by evaluation of the specific reactions that are planned for use in each synthesis, and each choice is examined to see which might be easier and better suited to available resources. This synthesis is based on the retrosyntheses in Figures 2.4 and 2.5, and is shown in Figure 2.6. The reagents are supplied from those listed in Chapter 8. Addition of a Grignard reagent (formed from bromobenzene, Reaction 50) to an acyl group, in this case an aldehyde (Reaction 56), converts **21** to **17**. Transformation of that alcohol to a bromide (Reaction 39) is followed by formation of the Grignard reagent (Reaction 50) and reaction with the epoxide (Reaction 55) to give **12**. Formation of the bromide from alcohol **12** (Reaction 39) and formation of the Grignard reagent (Reaction 50) is followed by addition of the Grignard reagent to propanal (Reaction 56). The final step is oxidation of the alcohol to a ketone (Reaction 09). Note that there are a total of nine steps (*not* counting the hydrolysis steps) but all reactions are straightforward and there should be no major problems.

CHAPTER 3

Computer-Assisted Syntheses

The previous chapters relied on a fundamental knowledge of organic reactions based on an undergraduate course, using Chapter 8 as the source of reagents for a given reaction. There are, of course, many more reagents, and vast areas of organic chemistry that are not represented in Chapter 8. There are many ways to approach the problem of finding and using new reagents. The most reliable approach is an excellent working knowledge of current literature, talking with colleagues, attending seminars and key meetings such as Gordon Research Conferences.

Nowadays, there are computer search engines that can be used in a variety of ways to scan the literature, but can also be used to guide retrosynthesis and synthesis. One of these computer-assisted methods is known as Reaxys, and it can be used to examine the literature for reagents and methods for a target, and/or for each disconnect transformation. Returning to the initially set problem of preparing **4** from **1** (see Figure 3.1), the retrosynthetic analysis can be done again, but this time using Reaxys to generate both the retrosynthetic pathway, as well as the reagents that are required.

This discussion will not go into the details of using the program itself, which is available by examining the Reaxys program itself. This discussion will assume the reader has access to Reaxys using their current web browser. Using the drawing tools provided by Reaxys, enter the first reaction based on the disconnection (pentan-2-one to 2-methylpentan-2-ol, without the numbers **3** or **4**).

After input, Reaxys returns 1 reaction out of 1 citation, and clicking on the "synthesize" link displays Figure 3.2, which is a reaction scheme along with the reference. This result obviously points to the Grignard reaction of methylmagnesium iodide with pentan-2-one, which validates this choice for a disconnection as described in Chapter 1.

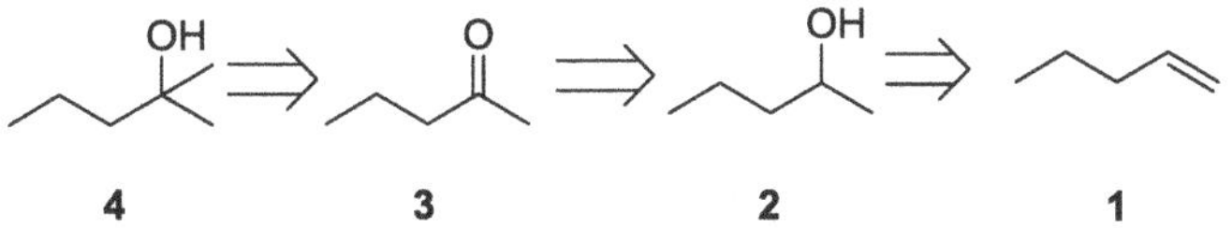

Figure 3.1 Retrosynthesis of 4.

Hybrid Retrosynthesis.
DOI: http://dx.doi.org/10.1016/B978-0-12-411498-2.00003-6

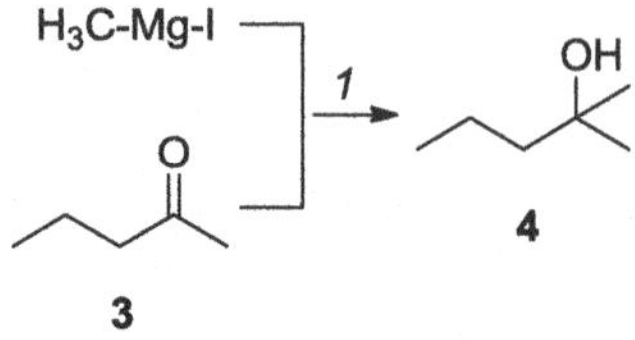

Step	Yield	Conditions	References
1		With diethyl ether	Van Rissenghem Bulletin des Societes Chimiques Belges, 1923, vol. 32, p. 149 *Chem. Zentralbl.*, 1923, vol. 94, #III, p. 1450

Figure 3.2 Reaxys display for the synthesis of **4**.

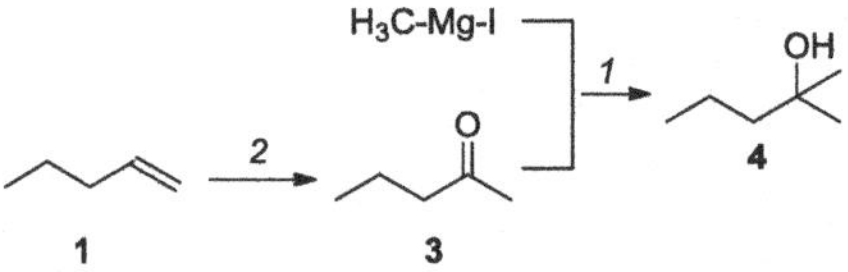

Step	Yield	Conditions	References
1		With diethyl ether	Van Rissenghem Bulletin des Societes Chimiques Belges, 1923, vol. 32, p. 149 *Chem. Zentralbl.*, 1923, vol. 94, #III, p. 1450
2	48%	With oxygen; N-hexadecyl-N,N,N-trimethylammonium bromide; copper dichloride; palladium dichloride in water;benzene T=80°C P=2327.2 Torr; Product distribution	Januskjiewicz, Krzystof; Alper, Howard *Tetrahedron Letters*, 1983, vol. 24, #47, p. 5159–5162
2		With oxygen palladium dichloride	Smidt et al. *Angewandte Chemie*, 159, vol. 71, p. 176, 180, 182 *Angewandte Chemie*, 1962, vol. 74, p. 93.
2	99% Chromat.	With perchloric acid; air; palladium diacetate; carbon in ethanol T=25°C; 4h	Cum, Giampietro; Gallo, Raffaele; Ipsale, Salvatore' Spadaro, Agatino *Journal of the Chemical Society*, Chemical Communications, 1985, #22, p. 1571–1573.

Figure 3.3 Reaxys generated synthesis of **4** from **2**.

One can ask if pentan-2-one (**3**) can be prepared from pentan-2-ol, or by a different route? Click on "synthesize" under pentan-2-one and 336 reactions out of 372 citations are returned, and the eighth example prepares pentan-2-one from pent-1-ene (**1**). When "add this reaction to plan" is clicked, Figure 3.3 is displayed, showing that **3** can be prepared directly from **1** using the reagents shown, and the references are provided. Completion of the synthesis of **4** using this approach is certainly a viable option.

There is an alternative route, based on the third reaction, using pentan-2-ol as the starting material. Clicking "add this reaction to plan" leads to Figure 3.4, and several references and reagents are shown that can be used to complete this transformation.

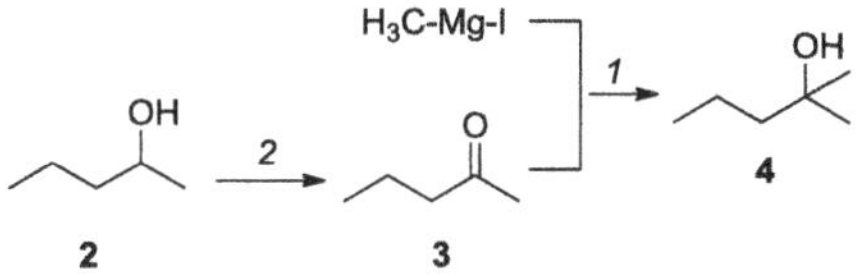

Figure 3.4 Reaxys synthesis of **4** from **2**.

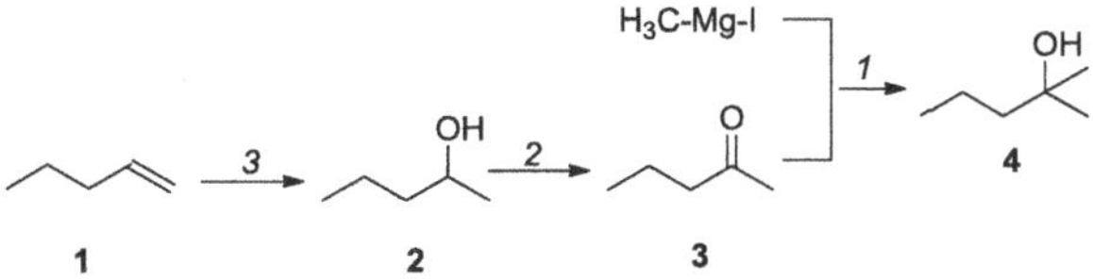

Step	Yield	Conditions	References
1		With diethyl ether	Van Rissenghem *Bulletin des Societes Chimiques Belges*, 1923, vol. 32, p. 149 *Chem. Zentralbl.*, 1923, vol. 94, #III, p. 1450
2	99%	With sodium dichromate; sulfuric acid; silica gel in dichloromethane T = 20°C; 5 h	Mirgalili, Bibi Fatemeh; Zolfigol, Mohamad Ali; Bamoniri, Abdolhamid; Zarei, Amin Phosphorous Sulfur and silico n and Related Elements, 2003, vol. 178, #8, p. 1845–1850
2	98%	With dihydrogen peroxide in water T = 89.94°C; 5 h	Ding, Yong; Zhao, Wei, Ma; Bao-chun; Qiu, Wen-yuan *Canadian Journal of Chemistry*, 2011, vol. 89, p. 13–18
2	97%	With pyridinium chlorochromate in chloroform T = 16°C; 168 h; or 1-methyl imidazolium chlorochromate or imidazolium chlorochromate	Agarwal, Seema' Tiwari, H.P. Sharma, J.P. *Tetrahedron*, 1990, vol. 46, #6, p. 1963–1974.

Figure 3.5 Reaxys generated synthesis of **4**.

Step	Yield	Conditions	References
1		With diethyl ether	Van Rissenghem *Bulletin des Societes Chimiques Belges*, 1923, vol. 32, p. 149 *Chem. Zentralbl.*, 1923, vol. 94, #III, p. 1450
2	99%	With sodium dichromate; sulfuric acid; silica gel in dichloromethane T = 20°C; 5 h	Mirgalili, Bibi Fatemeh; Zolfigol, Mohamad Ali; Bamoniri, Abdolhamid; Zarei, Amin Phosphorous Sulfur and silicon and Related Elements, 2003, vol. 178, #8, p. 1845–1850
2	98%	With dihydrogen peroxide in water T = 89.94°C; 5 h	Ding, Yong; Zhao, Wei, Ma; Bao-chun; Qiu, Wen-yuan *Canadian Journal of Chemistry*, 2011, vol. 89, p. 13–18
2	97%	With pyridinium chlorochromate in chloroform T = 16°C; 168 h; or 1-methyl imidazolium chlorochromate or imidazolium chlorochromate	Agarwal, Seema' Tiwari, H.P. Sharma, J.P. *Tetrahedron*, 1990, vol. 46, #6, p. 1963–1974.
3		Multi-step reaction with two steps 1. HI, 2 silver acetate/Verseifung des entstandenen Acetals durch Kali (hydrolysis of the resulting acetal by potassium carbonate)	Wurtz *Justus Liebig Annalen der Chemie*, 1868, vol. 148, p. 132

Figure 3.5 Reaxys generated synthesis of **4**.

To formally complete the initial retrosynthetic scheme, one can click on "synthesize" under pentan-2-ol, and several choices are given: 129 reactions out of 134 citations. One of them prepares **2** from **1**, and if one clicks "add this reaction to plan," the scheme in Figure 3.5 is presented. This scheme shows a complete synthesis, consistent with the original disconnection, along with references, reagents, and some information that concerns reaction conditions. One can obtain more detailed information by clicking on the appropriate boxes for each reaction, and even open the original reference if your organization has the correct access privileges.

This three-step sequence completes the analysis for the synthesis of **4** from **1**. Clicking on "output" or even "print" on the menu provided above the displayed results allows one to either print the screen or create a .pdf file that can be used later. This synthesis scheme is quite similar to that showed in Figure 1.9 in Chapter 1. A different oxidizing reagent is listed in this example (dihydrogen peroxide), which may be a new reagent for this transformation when compared to those presented in most classes. To be sure, learning new methodology is one advantage to using the Reaxys analysis. This analysis leads to alternative reagents, along with references for procedures, yield of products, etc. Using the library of reactions in Chapter 8 only gives general reagents and few details. The authors strongly recommend that primary literature be consulted for experimental protocols rather than following the descriptions found in Chapter 8. The conditions provided in Chapter 8 are meant to be a general, but representative guide.

Apart from a search using the target, **4**, it is possible to take structures **3** or **2**, one by one, and repeat this search process, but it is unnecessary since using the tools in Reaxys allows one to complete the planned retrosynthesis and also examine other possible routes (plan other retrosyntheses).

Based on results returned by Reaxys, the chemist doing the search will choose reagents and reaction conditions that are the most practical, and make the most sense to that person. In other words, the experience and knowledge of the chemist will be brought to bear to interpret the results. Some reactions returned by the search may use exotic reagents, or rely on such specific conditions that the reaction is impractical for the target being examined. In other cases, the reagents and/or conditions may be too harsh for the target, so deleterious side reactions are a problem. It is reasonable to assume that such reagents or such reaction conditions would be avoided by the chemist doing the search, but it is also possible that no alternatives are returned. In most cases, a computer search for any step will display a variety reactions and reagents. Such variety should provide several possible reagents and several possible synthetic routes. If not, the search may be modified.

One modification that should maximize the choices for a given target is to draw the target structure (**4**) and use Reaxys to search the target as the product. In a sample run using **4**, this action returned 54 reactions out of 79 citations. The option of multiple synthetic routes is available from this

Step	Yield	Conditions	References
1	90%	With sodium tert-pentoxide; sodium hydride; zinc(II) chloride in ethylene glycol dimethyl ether T=65°C; 12 h	Fort, Yves; Vanderesse, Regis; Caubere, Paul *Tetrahedron Letters*, 1985 vol. 26, #26, p. 3111–3114
2	98%%	With dihydrogen peroxide; *tert*-n-butylammonium in acetonitrile T=31.85°C; 10 h	Kanata, Keigo; Kotani, Miyuki' Yamaguchi, Kazuya; Hikichi, Shiro' Mizuno, Noritaka *Chemistry-A European Journal*, 2007, vol. 13, #2, p. 639–648
3	80%	With oxygen; isobutylaalaldehyde in 1,2-dichloromethane T–40°; 3 h	Kaneda, Kiyotmi; Hauna, Shigeru; Imanaka, Toshinobu; Hamamoto, Masatoshi; Nishiyama, Yutaka; Ishii, Yasutaka *Tetrahedron Letters*, 1992, vol. 33, #45, p. 6827–680
3		With 3,3-dimethyldioxirane in acetone T=25°C; Yield given	Baumstark, A.L.; McCloskey, C.J. *Tetrahedron Letters*, 1987, vol. 28, #29, p. 3311–3314.
3		With $Al_2(SO_4)$3-pumice stone T=300°C	Henne, Matuszak *Journal of the American Chemical Society*, 1944, vol. 66, p. 1651
4		With H_3PO_4-pumice stone T=250°C	Henne, Matuszak *Journal of the American Chemical Society*, 1944, vol. 66, p. 1651
4		With aluminum oxides; CuO/Cr_2O_3; potassium carbonate T=220°C; unter Wasserstoffdruck	Du Pont de Nemours ad Co. Patent US2004350, 1931;
4		With potassium hydroxide T=200–205°C	Guerbet *Comptes Rendus Hebdomadares des Seances de l'Academie des Sciences*, 1912, vol. 154, p. 225.
			Bulletin de la Societe Chimique de France, 1912, vol. (4)11, p. 283.

Figure 3.6 Reaxys generated synthesis of **4** from propan-2-ol.

action, so one is not bound to the original disconnection plan, but now have several options to consider. All of these routes cannot be presented here, but repeating this analysis will allow the reader to examine any and all possibilities. Choosing one route and then choosing "synthesize" for various intermediates stitches together the several different synthetic routes. One such route is shown in Figure 3.6, and by simply choosing "output" or "print," a .pdf file is assembled that can be downloaded, giving the reaction scheme along with the reaction conditions and the pertinent references.

It is clear from this analysis, that this alternative scheme is slightly longer than the original one proposed, but yields and cost/availability of starting materials and reagents, as well as those instances where multiple products are formed are now available for analysis. In the cases of multiple products, the issue of separation and purity of products must be considered. Apart from choosing the most efficient route to prepare **4**, it is also possible to use this exercise as a tutorial to examine other reactions, and perhaps learn some new chemistry.

Reaxys is not the only search computer-based engine that can be used for a search of chemical reactions relevant to a synthesis. SciFinder is one of the premier computer-based search engines that are available to an organic chemist. Using SciFinder Scholar® (SciFinder) for a substance search of **4** (from Figure 3.1) returned 31 hits. Careful inspection reveals that most of the hits are for the metal alkoxide salt of this compound.

When the cursor is placed on the alcohol, a >> icon is seen in the top right corner of the cell. Hovering over this icon and clicking opens a menu. One choice is "get commercial sources" since this substance is commercially available. This choice is a powerful, timesaving tool for the practicing chemist, though it is quite irrelevant to this generalized discussion of synthesis. Another option, far more germane to this discussion, is the button "synthesize this. . .". When selected, a total of 37 hits were returned. Long lists of hits are usually too time-consuming to search and shorter lists are more practical, so this result is a reasonable list of hits to examine. A variety of perfectly acceptable syntheses were returned including hydration of alkenes (Reaction 04 in Chapter 8), oxidations of alkanes and hydrolysis of an acetal (Reaction 08 in Chapter 8) just to name a few. One of the hits has the targeted final synthetic step (the first retrosynthetic step), shown in Figure 3.1. Checking the box to the left of the hit selects this hit. Scrolling back to the top of the page, near the right corner, one finds the button "send to SciPlanner." However, this SciPlanner appears to be a tool that organizes one's results/data. Thus, for continued searching, this function appears to be of limited utility, and will not be considered further in this discussion.

Returning to the selected hit, for **3**, it is possible to once again choose "commercial sources" or "synthesize this." Clicking on "synthesize this" returns a large number (575) of hits. Combined with the fact that the manner in which SciFinder displays the data is somewhat difficult to read, a complete examination of this large list is unmanageable. However, some refining can be done to make this search practical. On the left, the displayed set can be further analyzed in a number of ways. One is by reagent. Since the retrosynthesis in Figure 3.1 suggested a synthesis from pentan-2-ol, an oxidation reaction (Reaction 09 in Chapter 8) is required. Thus, oxidizing reagents should be examined first. Note that such a decision still requires some foreknowledge of chemical reactions on the part of the user.

An environmentally friendly preparation is often a desirable attribute, and, for the purpose of this example, NaOCl is an attractive alternative to the oxidizing agents used in Chapter 8. Selecting NaOCl and clicking "apply" reduced the number of hits to eight. For each of these hits, the user can immediately access the full text of the article, provided the institution has a subscription to SciFinder and the referenced journal. Aiming the cursor at the alcohol (which is now the starting material), the >> arrow is once again available to either search "commercial sources" or "synthesize." Choosing "synthesize this" once again returned a large and unmanageable number of hits (208). Since the previously discussed route is biased toward the hydration of pent-1-ene, hits can be restricted to reactions involving water (also see Reaction 04 in Chapter 8). This restriction reduced the number of hits to three, but adding an acid such as HCl increased the number of hits to 10, which is still a very manageable number. Checking for other acids such as H_2SO_4 did not greatly add to the number of hits. Curiously, reactions involving oxymercuration, which is a common method for generating alcohols from alkenes, are not listed. Furthermore, reactions that employ acidic ionic exchange resins or other immobilized acids are not immediately obvious.

In closing, either Reaxys or SciFinder can easily be used to navigate through a multistep synthesis. If, rather than having a targeted starting material as described above, there was no targeted starting material, one would likely stop the computer analysis once a familiar or commercially available precursor was reached via retrosynthetic analysis.

CHAPTER 4

A Hybrid Retrosynthesis Approach

Based on the Reaxys and SciFinder analysis of **4**(2-methylpentan-2-ol) in Chapter 3, it should be apparent that for relatively simple molecules, a complete synthesis can be devised without resorting to a complete retrosynthetic analysis. Indeed, if the target is entered and either search engine shows one or more synthetic routes, as with 2-methylpentan-2-ol, then one of those routes can be chosen. If the analysis of a relatively simple target such as 2-methylpentan-2-ol leads to multiple routes, then analysis of a more complex target such as **1A** may lead to a quite large number of possible routes. It is also possible that a computer search will not return any viable routes at all for a complex target, which often indicates that it has not previously been synthesized. For this reason, linking the use of Reaxys to the retrosynthetic analysis approach is deemed to be quite important. Searching for compounds that have not been previously synthesized is possible using SciFinder if the product has been reported as an isolated natural product. Although the "synthesize this" function will be unavailable if a compound has been reported, SciFinder will usually find that molecule, if the user enters the structure correctly. However, if the molecule has been synthesized, that route and those reactions will be available in the references.

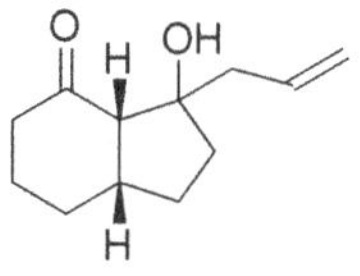

1A

In order to probe difficulties that may arise for the computer search of a more complicated target, compound **1A** has been chosen as a target. When structure **1A** was entered into Reaxys and a search was made using this structure as a product, the following was shown as a result of that search: "Reaxys could not find any hits in the database. Note that the query has

Hybrid Retrosynthesis.
DOI: http://dx.doi.org/10.1016/B978-0-12-411498-2.00004-8

been expanded automatically and a substructure search based on the original query also didn't find any hits." The parameters of the search were changed to: Search in reactions as "Any role." Search in substances and data. Create Alert from this query. Once again, the search gave: "Reaxys could not find any hits in the database. Note that the query was expanded automatically and a substructure search based on the original query also didn't find any hits." Based on this search alone, one could conclude that the molecule has not been previously synthesized. The next step is a retrosynthetic analysis, but can the computer be used to assist in this process?

If a literature search leads to the conclusion that a molecule such as **1A** has not been made, one could propose a *"hybrid" approach: disconnect step by step until a viable synthetic intermediate is found by a computer search that can generate a synthesis of that portion of the molecule. The chemist must "fill in the gaps" and devise a synthesis of the remainder of the molecule, based on the retrosynthetic analysis.* For example, one simple disconnection of **1A** removes the allylic group to generate the allylic disconnection fragment shown in Figure 4.1. Based on the bond polarization model in Chapter 1, the OH-bearing fragment is likely to be the acceptor fragment with **2A** as the synthetic equivalent. Therefore, the allylic fragment is likely the donor unit and correlates with the allylic Grignard reagent shown. This disconnection is now recognizable as an acyl addition reaction of a Grignard reagent to a ketone (Reaction 56 in Chapter 8). This conversion constitutes the last chemical step in the synthesis, so another disconnection of **2A** is required. To use the proposed hybrid approach, enter **2A** rather than **1A** as the search target in Reaxys.

A search returned 0 hits; and "Searching the query as 'substructure on heteroatoms'" also returned no hits. Reaxys automatically expanded the

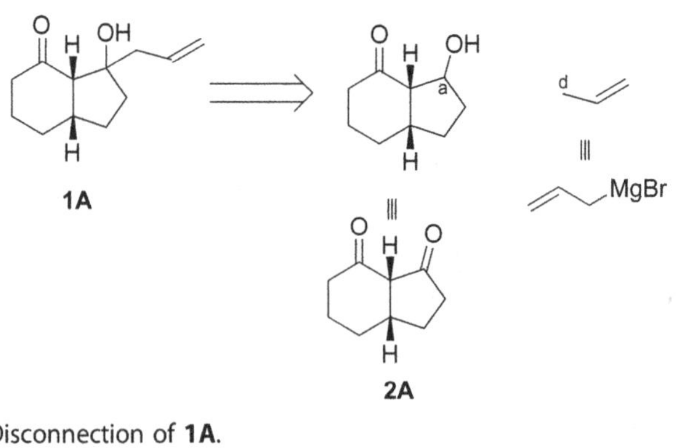

Figure 4.1 Disconnection of **1A**.

search to look for reactions related to the preparation of structurally related compounds, which led to one reference. The related compound and the accompanying reference showed that it was not close to **1A**, and although interesting, this result did not help. Repetition of the search with the "ignore stereo" box checked returned five reactions out of six substances and one citation was returned. Once again, the reactions shown in this citation do not help with the problem at hand. Another search after clicking the "find similar reactions" button returned no hits. Another search with the "no ring closures" box check return zero hits, as did the search in which the target was identified as "any role."

When SciFinder was used as the search engine, a search of **1A** as an "exact structure," no hits were returned. Another search that used "substructure" returned seven hits. SciFinder described five of these hits as having a relative stereochemical match and two lacking stereochemistry in the structures. Although this result has only a small number of hits to analyze, none appeared at all useful. The structure searched appeared as a part of far more complicated structures in various hits, to varying degrees. Thus, SciFinder does not appear to offer any significant advantages over Reaxys, but it does allow the user to execute the same type of computer-assisted planning. An alternative to the process described above using Reaxys may be to substitute another carbon-based group for the allyl group. This substitution (only for the sake of searching) requires the knowledge that an alkyl nucleophile would behave similarly to almost any other carbon-based nucleophile.

With that in mind, when **3** was searched for using the similarity search and not exact structure search, **4** was returned as a high-ranking hit. The two structures are close enough to each other that choosing synthesize this should allow the knowledgeable user to design a synthesis of **3** using appropriately modified intermediates or starting materials from the methods used to furnish **4**. Thus, although one can get quite far in a synthesis using such computer-based search engines, the bottom line remains that some knowledge must be applied to make use of the overwhelming amount of data available.

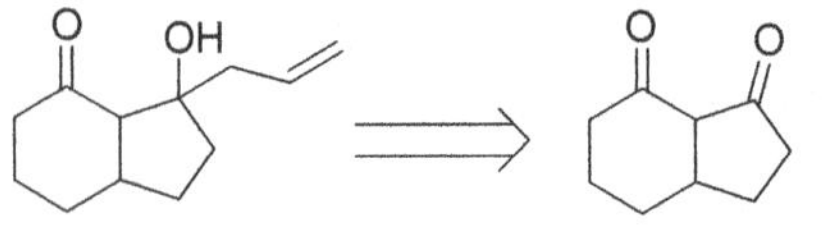

Figure 4.2 Disconnection of **1B** for a new computer search using **2B**.

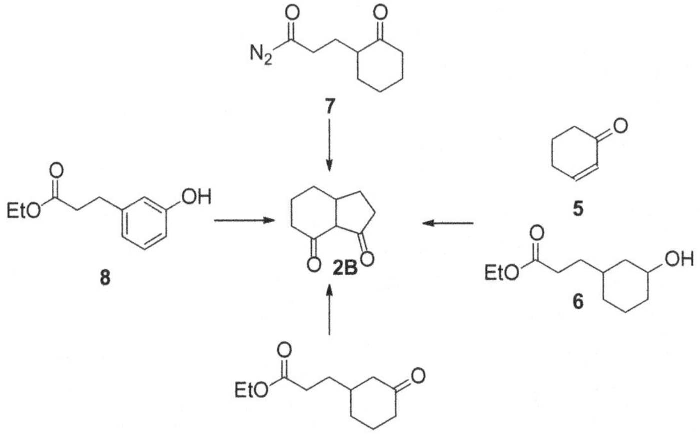

Figure 4.3 Reaxys generated synthetic precursors to **2B**.

One could try another disconnection, of course. However, note that the searches using **1A** and **2A** included specific stereochemistry for the stereogenic center. While it is not obvious, the results described above show that adding specific stereochemistry to **1A** and **2A** can limit the search using Reaxys. *In other words, if that specific compound is not recognized, possible "hits" may be overlooked.* Therefore, it is important to try the search again, but this time without specifying the ring juncture stereochemistry. SciFinder automatically performs searches with and without stereochemistry. As before, a search using Reaxys using **1B** returned no hits for that structure, but returned 92 reactions out of 17 citations for other structures and other reactions. While interesting, none seemed suited to the problem at hand. If **1B** is disconnected to the simpler structure **2B** (see Figure 4.2) a new search is possible using structure **2B**.

A Reaxys search of **2B** returned six hits, with five different references, all giving a synthesis of **2B**. Four starting materials were shown (see Figure 4.3), but two starting materials were shown for the same reference and three of the approaches seem very similar. At least two of the examples returned involve

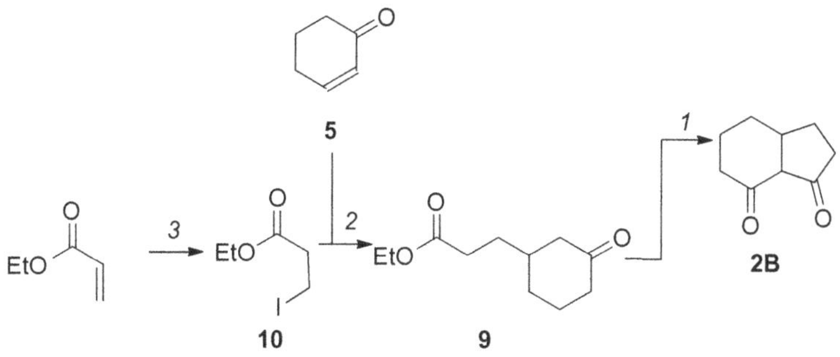

Figure 4.4 Reaxys generated synthesis of **2B** from ethyl acrylate.

what appears to be a conjugate addition (Reaction 78 in Chapter 8) to cyclohexenone (**5**). For this example, inspection of the reactions and references led to a choice of **6** as the precursor.[1] Expansion of this choice led to the sequence shown in Figure 4.4.

While these results do not necessarily solve the complete synthesis problem, they lead to several synthetic routes that might be applied, based on the first disconnection to **2B**. The conversion of **2B** to **1A** remains a problem that must be solved by another method. However, choosing one route to **2B**, and clicking "add to the plan" and then "synthesis" for each choice, led the reaction scheme in Figure 4.4.

> Note that this approach is a "hybrid" in that the initial target was not found using Reaxys, but one disconnection led to viable reactions, allowing one to complete the retrosynthesis, in this case leading to ethyl acrylate as a starting material.

The starting material in this sequence was ethyl acrylate, which was treated with HI in two of the references to give **10**. Subsequent reaction with zinc and Me_3SiCl, followed by treatment with copper diacetate and then Me_3SiCl, gave an organometallic that reacted with cyclohexenone (**5**) via conjugate addition (Reaction 78 in Chapter 8) to give **9**. Treatment with sodium metal in ethanol gave the enolate anion of the ketone, and intramolecular cyclization with the pedant ester (see Reactions 73 and 75 in Chapter 8) gave **2B**. Reaction 1 in Figure 4.4

---

[1] Stetter, H.; Krüger-Hansen, I.; Rizk, M. *Chemische Berichte*, **1961**, *94*, 2702−2707.

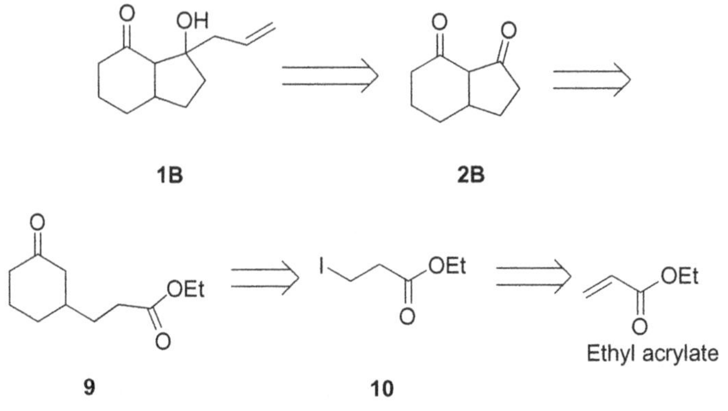

Figure 4.5 Hybrid Retrosynthesis of **1B**.

was taken from Stetter's work,[1] and there were three references for Reaction 2, which proceeded in 75%[2] or 50%[3] yield depending on the method (reactions conditions were shown in the output). Conversion of ethyl acrylate to **10** in step 3 was displayed with three references.[4] The synthesis shown in Figure 4.4 shows conditions from Refs.[5,3,6]

The *hybrid retrosynthesis process* of standard disconnection followed by a Reaxys search of disconnect intermediates led to the synthesis shown in Figure 4.4, with ethyl acrylate as the starting material. To establish the stereochemistry, one must refer to the literature to ascertain if it is possible to control the various reactions to give the desired *cis*-ring juncture in **1B**. Note that a search using Reaxys did not solve the first disconnection of **1B** to **2B**, but the disconnection was made with the recognition that a Grignard reaction could be used for that step. This choice was made using the classical retrosynthetic approach, with knowledge of pertinent chemical reactions (Figure 4.5).

[2] (a) Hill, C.L.; McGrath, M.; Hunt, T.; Grogan, G. *Synlett.* **2006**, 309−311; (b) Hu, Y.; Yu, J.; Yang, S.; Wan, J.-X.; Yin, Y. *Synth. Commun.* **1998**, *28*, 2793−2800.

[3] Baker, F.W.; Parish, R.C.; Stock, L.M. *J. Am. Chem. Soc.* **1967**, *89*, 5677.

[4] (a) Caps, P.J.; Garcia, B.; Rodriguez, M.A. *Tetrahedron Lett.* **2002**, *43*, 6111−6112; (b) Moureu, M.; Tampier, *Annules de Chimie (Cachan, France)*, **1921**, *9(15)*, 246; *Compt. Rend. Hebd. des Seances de 'Alacemie des Sci.* **1921**, *172*, 1269; (c) McCaffery, E.L.; Shalaby, S.W. *J. Organomet. Chem.* **1972**, *44*, 227−231.

[5] Holtz, H.D.; Stock, L.M. *J. Am. Chem. Soc.* **1964**, *86*, 518.

[6] Seebach, D. *Angew. Chem. Int. Ed.* **1979**, *18*, 239.

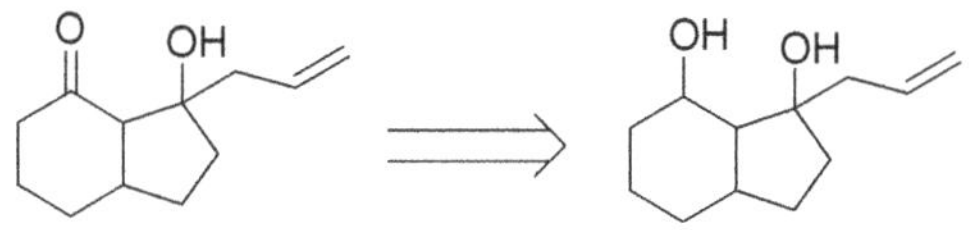

Figure 4.6 Functional group exchange of **1B** to **11**.

There is a problem in this retrosynthesis! The Grignard reaction of an allylic Grignard reagent to a ketone appears to be straightforward, but there are two ketone moieties. There is no reason to believe that the reaction would be selective for one carbonyl unit in preference to the other, so at least two different products should result.Remember that no synthesis of **1B** could be found. It is left to the reader to evaluate the efficacy of a Grignard reaction of **2B** and an allylic Grignard reagent to give **1B**. The inability to find a solution to this problem using the data from Reaxys, as presented so far, is sufficient to force a reevaluation of this synthesis. In other words, structural changes are required to the synthetic intermediates. Note that this conclusion was reached using knowledge of chemical reactions, but the fact that there were no "hits" for conversion of **2B** to **1B** certainly suggested there was a problem (Figure 4.6).

One option is to modify the starting material **1B** to generate **11** by a simple functional group change. If **11** can be prepared, a simple oxidation would give **1B**. Using Reaxys with **11** as the search target returned 0 hits, but the automatic searching capability for related structures returned 137 reactions with 28 citations. Scanning these possibilities revealed three interesting transformations, and checking those items from the original list of 137 and then clicking on "limiting the selection" led to a closer examination of those three items. Two of the interesting reactions suggest that it is possible to modify a ketone moiety in the presence of an alcohol moiety. The third item suggests that one carbonyl group can be protected as adioxolane,[7] allowing selective Grignard reaction at the second carbonyl. Based on these examples, which do not actually generate the targets of interest, a different idea for a synthesis can be generated, using a new disconnection of **1B**, to **12**. While protecting groups are not

[7] Wuts, P.G.M.; Greene, T.W. *Greene's Protective Groups in Organic Synthesis* 4th ed., Wiley, NJ, *2006*, pp 306−318.

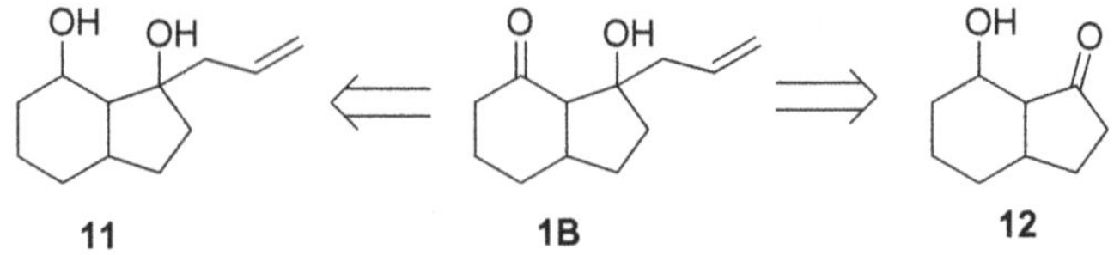

Figure 4.7 Two disconnections of **1B**.

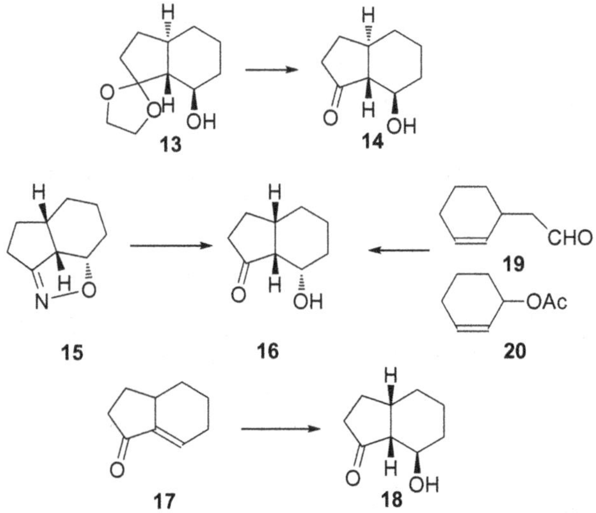

Figure 4.8 Reaxys generated synthesis of diastereomers of **12**.

discussed in this book, a working knowledge of protecting groups is often essential for synthetic planning[8] (Figure 4.7).

Using **12** as the search target for Reaxys returned 39 reactions with four citations. Scanning through the possibilities led to limiting the output to four reactions in three citations, as shown in Figure 4.8. In all cases, racemic products were prepared, but note that the stereochemistry of the three stereocenters is defined in all reactions returned by Reaxys, so these sequences are diastereoselective. The use of **13** involves the

---

[8] (a) Greene, T.W. *Protective Groups in Organic Synthesis* Wiley, NY, *1980*; (b) Wuts, P.G.M.; Greene, T.W. *Protective Groups in Organic Synthesis* 2nd ed., Wiley, NY, *1991*; (c) Wuts, P.G.M.; Greene, T.W. *Protective Groups in Organic Synthesis* 3rd ed., Wiley, NY, *1999*; (d) Wuts, P.G.M.; Greene, T.W. *Greene's Protective Groups in Organic Synthesis* 4th ed., Wiley, NJ, *2006*; (e) Wuts, P.G.M. *Greene's Protective Groups in Organic Synthesis* 5th ed., Wiley, NJ, *2014*. Also see Smith, M.B. *Organic Synthesis,* 3rd ed., Wavefunction/Elsevier, Irvine, CA/Waltham, MA, pp. 587−622.

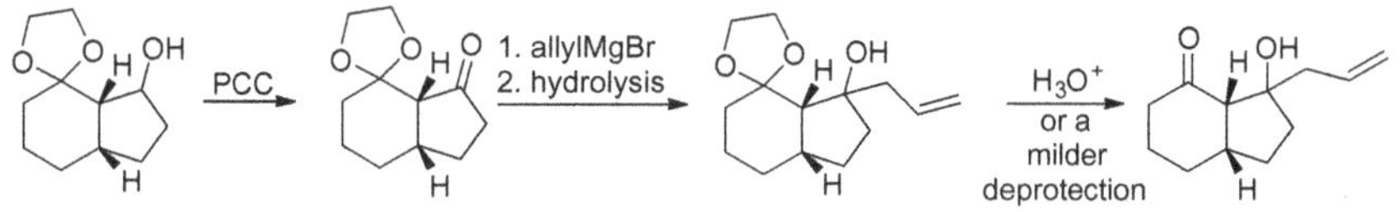

Figure 4.9 Synthesis of **1A** from **21**.

aforementioned dioxolane-protecting group to give diastereomer **14**.[9] Formation of the oxazoline species **15** is required for the second approach, which gives **16**.[2,10] The Reaxys search suggested syntheses of diastereomer **16** from aldehyde **19** or ester **20**.[4] Note that the *cis*-ring juncture in the original structure **1A** has been generated in **16**. Conjugated bicyclic ketone **17** was used in the last approach to give diastereomer **18**,[4,2] which also has the correct *cis*-ring juncture for **1A**. Note that **16** and **18** have a *cis*-ring juncture, but different stereochemistry from the hydroxyl-bearing carbon atom. However, this carbon will eventually be converted to a ketone unit, so the stereochemistry at that site is unimportant, except for purification of that intermediate compound. The same references and reactions are returned for an identical search using SciFinder, rather than Reaxys. On some level, this observation should provide relief to the user since it means that either search engine can be used with the expectation of the same or similar results. With these results in hand, a synthesis to the target should be straightforward.

Choosing one possible transformation, say **14** from **13**, allows one to use Reaxys and then click on "synthesize" for each structure, which leads to several possible syntheses. This approach does not necessarily give a perfect solution, and it requires chemical knowledge of the various possibilities. Availability of materials and reagents, and viability of each procedure must be evaluated. Once again, the computer search leads to possibilities, but a hybrid approach is required using retrosynthesis, computer searching, and knowledge of reactions to construct the final synthesis.

The target **1B** has a *cis*-ring juncture, but all the hits shown have a *trans*-ring juncture. However, using the hit as a guide, assume that it is possible to modify the approach suggested by returned results to prepare the related compound **21**. Assuming that **22** is a precursor to intermediate **18**, as shown in Figure 4.9, this modification offers a possible solution to this portion of the synthesis. Presumably, a synthesis of **17** can be

---

[9] (a) Tsantali, G.G.; Dimtsas, J.; Tsoleridis, C.A.; Takakis, I.M. *Eur. J. Org. Chem.*, **2007**, 258-265; (b) Wollenberg, R.H.; Goldstein, J.E. *Synthesis* **1980**, 757–758.

[10] Curran, D.P.; Jacobs, P.B. *Tetrahedron Lett.***1985**, *26*, 2031–2034.

modified to incorporate the dioxolane unit in **22**, and a retrosynthesis of **17** is shown because the hybrid retrosynthesis using Reaxys generated a complete reaction scheme that used 4-chlorobutanol as the starting material. Once **21** is in hand (Figure 4.9), oxidation of the alcohol moiety (Reaction 09 in Chapter 8) will give the ketone moiety in **22**, which allows reaction with allylmagnesium bromide (Reaction 56 in Chapter 8) to give **23**, and deprotection of the dioxolane unit (Reaction 08 in Chapter 8) finally leads to **1A**. Note also, that this modified or hybrid approach leads to the originally targeted *cis*-ring juncture.

This example generated a synthesis of **1A**, but more importantly, it provided an example that how a search is conducted is vitally important. An important lesson is that one must use chemical intuition and evaluation of any search must be practical. Even taking the time saving power of computer searches into account, there is no replacement for a good working knowledge of reactions and selectivity. A computer search can point the way, but the chemist must intervene to interpret the myriad of results, especially with the hybrid retrosynthesis approach. The hybrid combination of retrosynthetic analysis and using both Reaxys and SciFinder to screen possible reactions relied on the following protocol:

- Do a search with and without stereochemistry (SciFinder appears to do this automatically).
- If hits are not found, do a disconnection and search the disconnect product.
- Continue this process until Reaxys can provide reasonable routes.
- Use the literature citations to search for solutions to stereochemistry issues.
- Use the literature citations to give alternative routes. They may even suggest more interesting alternatives.
- Take care to analyze the results, looking for compatibility issues, availability of reagents/compounds, and the possibility of generating related by alternative targets such as **2B** where selectivity is a serious problem.

The synthesis of **4** and of **1A** used carbon−carbon bond-forming reactions as the key step, and the focus of the disconnection approach clearly centered on those reactions. However, it is also clear that most of the chemical transformations in both syntheses simply manipulated functional groups, either because certain functionality was required to make the carbon−carbon bond, or because that functional group was present in the final target. This observation raises the issue of using computer-based searches to find not only carbon−carbon bond-forming reactions but also functional group interchange reactions.

CHAPTER 5

Creative Strategies to Searching for Reagents

As noted in previous chapters, it is reasonable to assume that most reactions from an undergraduate organic chemistry course involve polarized covalent bonds and/or ionic intermediates. With such an assumption, most reactions involve functional group manipulation, and the bonds connecting carbon to a functional group are examined in an effort to find disconnections that simplify the target. In previous discussions, the reagents to be used in a synthesis were supplied either by examination of reactions presented in Chapter 8, or were taken from the literature references supplied by a computer search, whether it was from Reaxys or SciFinder. As a practical matter, the generic reagents listed in Chapter 8 may not work for every substrate, or there may be reactivity and/or isolation problems associated with their use. In principle, Reaxys will generate a synthetic route for a given target, along with references that suggest reagents for each transformation. The use of these reagents may solve the problem of how to accomplish each transformation. As seen in the hybrid retrosynthetic approach used for **1A** in Chapter 4, there may be steps for which no known reagents and conditions have been reported. Finding a reagent is therefore rather important, and the original literature is the only useful source.

Is it possible to generate a longer and more detailed list of reagents for each of these transformations using Reaxys to scan the literature? The answer is yes! New reagents or certainly alternative reagents can be found using rather simple molecules in the search engine. This approach is not perfect, of course, and will not scan the literature for all possibilities. Only diligent reading of the literature will give that information, but alternatives will certainly be found to expand the library of reactions at one's disposal. The following examples are based on a Reaxys search for specific examples of the 12 reactions given in Chapter 2. Note that in all 12 cases, choosing different reactants and products can and probably will lead to a different list, although many redundant citations are expected.

Hybrid Retrosynthesis.
DOI: http://dx.doi.org/10.1016/B978-0-12-411498-2.00005-X

It is probably important to do searches with a few different transformations to get a more realistic sense of what may be available.

1. **S_N2 reactions of alkyl halides or alkyl sulfonate esters with cyanide ion or alkyne anions.**

   In Chapter 8, reagents used for Reactions 67 and 81 are relevant to this transformation. Using Reaxys, there was a search for the transformation of 2-bromopentane to 2-methylpentanenitrile and of 2-bromopentane to 3-methylhex-1-yne. This search was used to illustrate C—C bond-forming reactions and also give examples to use as a prototype and reagents that can be considered for the transformation.

   The nitrile search produced one hit. The alkyne search produced 0 hits, but the automatic expanded search generated 651 reactions out of 192 citations.

   With SciFinder, the nitrile returned zero hits, while the automatic expanded search gave 2363 reactions. Meanwhile, the alkyne search also returned zero hits but the substructure search gave 1955 reactions. SciFinder allows for very easy analysis of the results and sorting. For example, some of the more useful options (although there are others) are sorting by: document type (i.e. patent, report, journal, etc.); language; number of steps; product yield; reagent; and solvent. Such analysis and refining can be done for any reaction on SciFinder, including all of those listed below.

2. **S_N2 reactions of enolate anions (from ketones, aldehydes, esters) with alkyl halides or alkyl sulfonates.**

   In Chapter 8, reagents used for Reactions 58 and 74 are relevant to this transformation. Using Reaxys, the search focused on the transformation of butan-2-one to pentan-3-one. The search returned zero hits, but the automatic expanded search returned 1,205,257 reactions with of 149,189 citations, which is an unmanageable number of hits to evaluate. The structure was changed and 2-methylcyclohexanone from cyclohexanone was searched in order to satisfy the transformation but limit the possibilities. This new search returned 17 reactions with 16 citations, which is a far more manageable number of hits.

   When performing this search with SciFinder, seven reactions are returned, though none appear to be a traditional enolate alkylation. When the search is expanded as a substructure search, in excess of 2. 5 million reactions are returned. This is a clearly a laughable number of hits. In this case, the number of hits is also unworkable when using the refining protocols; the number of hits is too large. This result means that searching for a general procedure to use for an enolate alkylation is not a viable approach, and some specific reaction must be searched. For example, if the conversion of acetophenone to ethyl phenyl ketone is searched, a workable 18 hits are returned. An analysis of the reagents reveals the use of bases such as KH and K_2CO_3 in several cases, a telltale sign for enolate alkylation, which could not be recognized without at least some familiarity with organic reactions.

3. **Substitution reactions of organometallics such as organocuprates with alkyl halides or alkyl sulfonates.**

In Chapter 8, reagents used for Reaction 62 are relevant to this transformation. The search for the reaction of lithium dimethyl cuprate and 1-bromopentane using Reaxys was the goal, but to make Me_2CuLi, one must use the periodic table button and choose Cu and Li for the appropriate positions. How one enters the structure for Me_2CuLi could cause problems in the search. This particular search returned three reactions with three citations, but they involved much more complex structures. The structures were changed to a reaction of 1-iodooctane with lithium dimethyl cuprate to give nonane, but this search returned no hits. If the Me_2CuLi structure was removed so the new search was for bromooctane to nonane, Reaxys returned 21 reactions with 15 citations, and although no Gilman type reagents (organocuprates) were shown. If the search was modified again to convert a pentyl derivative with any metal (M) to decane, the Reaxys search returned 349 reactions out of 486 substances and 162 citations. This time, some examples involves organocuprates, but the entire list must be examined. If the search is done again, but limited to reagents/ catalyst contains cuprate, 69 reactions out of 86 substances and 15 citations were returned. Once again, this list must be searched, but it is clear that the choice of proper limiters is important for this type of search. This brief discussion points out a difficulty in finding such a match, at least with this particular search. Similar difficulties exist when searching for other reactions as well.

With SciFinder, when the conversion of 1-bromobutane to hexane is queried, zero hits were returned. If the functional group icon is chosen, a search of 1 bromobutane with any organometallic reagent can be performed. This search returns only 2147 hits, including at least several using cuprates, including the addition of a variety of alkyl groups.

4. **Acyl addition of organometallics such as Grignard reagents and organolithium reagents with aldehydes or ketones**.

In Chapter 8, reagents used for Reactions 56 and 68 are relevant to this transformation. The search for an acyl addition reaction used the conversion of cyclopentanone to methylcyclopentanol. This Reaxys search returned 15 reactions out of 24 citations.

The same search with SciFinder gave 11 reactions. A variety of different conditions are returned, including the expected traditional Gilman reaction of methyllithium with methyl Grignard reagents. A methyl trialkoxyltitanium reagent is also returned. Thus, sometimes, less well-known conditions can be found using this process.

5. **Acyl addition of enolate anions (from ketones, aldehydes, esters) to aldehydes or ketones; the aldol condensation.**

In Chapter 8, reagents used for Reactions 57, 80, and 73 are relevant to this transformation. The search for this reaction type used the condensation of butanal with cyclopentanone. This Reaxys search returned nine reactions out of six citations, and one was an aldol condensation.

On SciFinder, rather than probe a specific reaction between cyclopentanone and butanal, the two were simply entered in as reagents. Using this strategy, 32 reactions were returned. Most of the reactions were aldol condensations, like the search with Reaxys. Other reactions were returned, but they are more complicated although at some point both reagents are involved. This return of additional and presumably less common reactions undoubtedly may complicate matters, but the small number of reactions returned (at least in this case) is easy enough to navigate.

6. **Acyl substitution such as the Claisen condensation with esters.**

In Chapter 8, reagents used for Reactions 73 and 75 are relevant to this transformation. The search for this reaction type used the condensation of ethyl butanoate with ethyl acetate. This Reaxys search returned three reactions with four citations. The first citation was a Claisen condensation that used Na to generate the enolate anion.

The same strategy mentioned earlier was used in this example with SciFinder and 10 hits were returned, all of which were completely irrelevant. However, if the search was changed to use ethyl benzoate and ethyl butyrate as the reacting partners, five hits are returned but only one is an obvious Claisen condensation.

7. **Conjugate addition of organocuprates to conjugated aldehydes, ketones, or esters.**

In Chapter 8, reagents used for Reaction 78 are relevant to this transformation. The search for a conjugate addition avoided the use of cuprates but used the conversion of 2-cyclohexenone to 3-methylcyclohexanone. This Reaxys search returned 74 reactions with 74 citations. One of them involved the specified organocuprate addition. If the search was repeated but the limiter of reagents/catalysts contains cuprate was employed, the search returned four reactions out of eight substances and 10 citations.

With SciFinder, searching for the conversion with no limiters returned 202 reactions. Scrolling through the results reveals the use of a variety of organometallics including the expected cuprates. The search can be refined easily as described earlier to employ a variety of copper salts such as CuBr, CuI, and CuCN.

8. **Friedel-Crafts alkylation and acylation with aromatic substrates.**

In Chapter 8, reagents used for Reactions 69 and 77 are relevant to this transformation. Although the Friedel-Crafts acylation reaction is probably more useful synthetically, both constitute a $C-C$ bond-forming reaction. Beginning the Reaxys search with the reaction of benzene to give diphenylmethane. This search returned 267 reactions out of 251 citations. One of the hits involved the reaction of benzene with benzyl chloride in the presence of the Lewis acid, aluminum chloride. The Reaxys search, which used the reaction of benzene with ethanoyl chloride (acetyl chloride), returned 204 reactions out of 193 citations.

When benzene was used as the starting material in a search to give diphenylmethane, using SciFinder, 476 hits were returned. According to the analyze window, 23 of the reactions use $AlCl_3$, a known Lewis acid for Friedel-Crafts alkylation reactions. With the benzene/acetyl chloride reaction, 1147 hits were returned. Once again, hits in addition to the desired Friedel-Crafts alkylation reaction may be returned due to the way the search was executed. For example, the synthesis of 2,4-dichloroacetophenone is returned, despite not allowing for variable structures. If the reagent list for this reaction is examined closely, although the reaction figure starts with benzene and acetyl chloride, step 1 is an aromatic chlorination, followed by the Friedel-Crafts acylation. A general search like this, where only reagents are entered into the search, can be useful, but a specific targeted reaction may be harder to find. That being said, in this case, the number of hits returned was small enough to enable refining and analyze to be active. The link to the 418 reactions using $AlCl_3$ could easily be selected.

9. **Wittig reactions with aldehydes or ketones to give alkenes.**

In Chapter 8, reagents used for Reaction 66 are relevant to this transformation. The Reaxys search used the conversion of cyclohexanone to methylenecyclohexane. This search returned 24 reactions out of 22 citations. Several simple examples of the Wittig reaction were returned, along with other reagents for this transformation.

With SciFinder, 79 reactions were returned when the same search was performed. Even using the analyze/refine function, only one reaction obviously was the expected Wittig reaction. However, several reactions use bases, which are required in a Wittig reaction. Selecting these, it becomes clear that these too are Wittig reactions since the reaction figure shows that an alkyl phosphonium salt is one of the reagents. Once again, even when searching reactions with a very specific starting material and product some level of oversight by the chemist is necessary and this is only possible with a strong foundation in chemical reactivity.

10. **Diels–Alder cyclization of dienes with alkenes.**

In Chapter 8, reagents used for Reaction 64 are relevant to this transformation. This reaction is arguably one of the more powerful chemical reactions in organic chemistry. A simple Reaxys search used the reaction of cyclopentadiene with methyl vinyl ketone. This search returned 16 reactions out of 86 citations. If the search was repeated using the conversion of 1,3-butadiene to cyclohexene, but with a generic group (G) at the C4 position relative to the C=C unit of the cyclohexene, the Reaxys search returned 213 reactions out of 371 substances and 228 citations, which may be a more useful search in this particular case.

With SciFinder, 208 reactions were returned with this search. Most of the changes appear to involve additives or solvents that may or may not influence the rate of cycloaddition. In at least one case, the stereochemistry of the product is different, according to the reaction figures. This result would be very useful to the user if this stereoisomer was the one desired. It is not clear, however, if this change is the typical stereochemistry observed in other reactions, or just this one. The answer can only be determined by obtaining the original article.

11. **Sigmatropic rearrangements such as the Cope and Claisen rearrangement.**

In Chapter 8, reagents used for Reactions 65 and 72 are relevant to this transformation. The Claisen type rearrangement can be probed by searching the reaction that converts allyl vinyl ether to pent-4-enal, making the stereochemistry of the product a mixture of E and Z isomers by using the squiggle line shown. This search returned one reaction out of three citations. The first example shown is a typical Claisen condensation. Search a Cope type rearrangement used the conversion of hexa-1,5-diene, with a generic group (G) at C3 to give a 1,5-diene was the G group attached to one C=C unit all with the "squiggle" line to indicate a mixture of stereoisomers. This Reaxys search returned 42 reactions out of 73 substances and 17 citations. If a new search converted hexa-1,5-diene to hexa-1,5-diene, but inserted a G group at C3 in both structures, the Reaxys search returned 32,077 reactions out of 32,267 substances and 7106 citations.

Searching for the same pentenal reaction using SciFinder, 20 hits were returned. The reactions that are sigmatropic rearrangements just use heat and time. At this point, it is worth pointing out the "similar reactions" button provided for each hit. Clicking this button allows for a new search to be done that either just looks at the answer set or at all reactions for a broad (reaction centers only) search, medium (reaction centers and adjacent atoms and bonds) search, and narrow (reaction centers and extended atoms and bonds) search. Performing a broad search in this case that looks at all reactions returned 57 hits, including one that furnishes 3-phenylpent-4-enal *via* a Claisen rearrangement. This appears to be a more directed substructure search that vastly reduces the number of hits the substructure searches typically return.

12. **Transition metal-catalyzed (i.e., Pd, Ru) C−C bond-forming reactions.**

In Chapter 8, reagents used for Reactions 63, 70, and 65 are relevant to this transformation. A Reaxys search must focus on the specific transformation, using specific simple examples of the Heck reaction, ring closing metathesis, the Sonogashira reaction, and the Suzuki-Miyaura coupling. A search for the conversion of methyl acrylate to methyl 3-phenylacrylate, using the "squiggle line" to indicate a mixture of E/Z isomers returned 115 reactions out of 107 substances and 370 citations. A search using the conversion of octa-1,7-diene to give cyclohexene returned 5 reactions out of 10 substances and 22 citations. A search for the conversion of halobenzene (a generic halogen, X) to biphenyl returned 541 reactions out of 658 substances and 1278 citations. Finally, a search using the conversion of halobenzene (a generic halogen, X) to 1-phenylpropyl-1-yne returned five reactions out of nine substances and 5 citations. It is certain that additional searches using more variation of starting materials, perhaps with some limiters, will generate many examples of new reactions and reagents for consideration.

On SciFinder, with the first case, 968 hits were returned. Of these, only four obviously used Pd when examined via the "analyze" function. Only the (E) isomer was searched for. In the second reaction, this search returned 63 hits. However, none of the reagents listed are obviously a transition metal catalyst. The results show rather odd number abbreviations for reagents. The user will need to read the title of the reference (which is clearly provided under the reaction figure) to determine which catalyst is employed. In the third case, using SciFinder to search the transformation of chlorobenzene to give biphenyl gave 850 hits. Some of the first hits use a Pd catalyst and phenyl boronic acid as the reacting partner. In the final reaction, once again using chlorobenzene to give 1-phenylpropyl-1-yne, no hits were returned. If the Cl is changed to a Br, two hits are returned. One of these hits employed palladium-catalyzed coupling between the bromobenzene and an organozinc.

There are, of course other reactions that generate carbon−carbon bonds. Any modern graduate course in organic chemistry will greatly expand the list of possibilities. The intent of this exercise is to provide an approach to supplement the library of reactions that most chemists will have after an undergraduate course.

Exactly the same approach may be used to supplement reagents and examples for various Functional group exchange reactions. As first introduced in Figure 1.7 in Chapter 1, Figure 5.1[1] provides a visual reminder of the relationship of many functional groups. Note that while Figure 5.1 is a reproduction of Figure 1.7, reactions from Chapter 8 are supplied for the various transformations as the numbers in parentheses following each reaction *a-x*. It is possible to choose specific transformations for relatively simple molecules to provide an expanded list of reagents. Search results will be provided given for each reaction in Figure 5.1 using Reaxys or SciFinder.

[1] Smith, M.B. *J. Chem. Ed.* **1990**, *67*, 848−856.

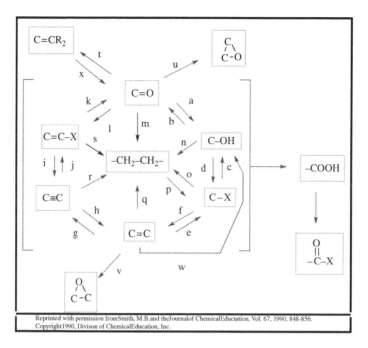

Reprinted with permission from Smith, M.B. and the Journal of Chemical Eductation, Vol. 67, 1990, 848-856. Copyright 1990, Divison of Chemical Education, Inc.

a. butan-2-one to butan-2-ol (06)

b. butan-2-ol to 2-butan-2-one (09)

c. butan-2-ol to 2-bromobutane (39)

d. 2-bromobutane to butan-2-ol (see 41 and 04)

e. but-2-ene to 2-bromobutane (41)

f. 2-bromobutane to but-2-ene (18)

g. but-2-ene to but-2-yne (see 41 and 19)

h. but-2-yne to but-2-ene (17)

i. but-2-yne to 2-bromobut-2-ene (43)

j. 2-bromobut-2-ene to but-2-yne (19)

k. 2-bromobut-2-ene to butan-2-one (45)

l. butan-2-one to 2-bromo-but-2-ene (not listed)

m. butan-2-one to butane (16)

n. butan-2-ol to butane (04)

o. 2-bromobutane to butane (15)

p. butane to 2-bromobutane (40)

q. but-2-ene to butane (14)

r. but-2-yne to butane (see 17 and 14)

s. 2-bromobut-2-ene to butane (14)

t. butan-2-one to 2-methyl-but-1-ene (66)

u. butan-2-one to 2-ethyl-2-methyloxirane (not listed)

v. but-2-ene to 2,3-dimethyloxirane (34)

w. but-2-ene to butan-2-ol (04)

x. 2-methylbut-1-ene to butan-2-one (10)

Figure 5.1 Functional group exchange reaction wheel. A visual reminder of the chemical relationship of functional groups. The numbers in parentheses, (04), refer to the pertinent reactions in Chapter 8. *Reprinted with permission from Smith, M.B. and the* Journal of Chemical Education, *Vol. 67, 1990, 848–856.Copyright 1990, Division of Chemical Education, Inc.*

a. A search of the reaction butan-2-one to butan-2-ol returned 38 reactions out of 104 citations. The use of $LiAlH_4$ and also $NaBH_4$ is rather extensive, but many more reagents are shown for this transformation.

With SciFinder, 212 hits were returned. It is not clear why there was such a difference in the searches. Once again, the reagents $NaBH_4$ and $LiAlH_4$ were reported. Other conditions were also returned, including hydrogenation and what appears to be some stereoselective catalysts. However, SciFinder did not allow for a stereospecific search.

Also see Reaction 06 in Chapter 8.

b. A search of the reaction butan-2-ol to butan-2-one returned 64 reactions out of 238 citations. Many examples illustrate a fundamental oxidation method using Cr(VI).

Using SciFinder, 434 hits were returned. Not surprisingly, a variety of conditions exist for this transformation, including the expected Cr (VI) reagents. Other reagents, such as bleach (NaOCl), N-methylmorpholineoxime, $NaIO_4$, and $KMnO_4$ are also easily found with the "analysis" functionality.

Also see reaction 09 in Chapter 8.

c. The search for butan-2-ol to 2-bromobutane returned 13 reactions out of 22 citations. Many different brominating reagents are shown. If the search is expanded to target a generic halide (X) rather than bromine, the search returned 30 reactions out of 36 substances and 53 citations.

Using SciFinder, the same search found 10 reactions. One reaction is clearly listed as an S_N2 reaction in the reaction figure. Several different expected reagents exist such as PBr_3 and HBr. Also, conditions are returned that are less obvious to the inexperienced chemist by simply looking at the" analyze reagents" window. For example, H_2SO_4 is listed as a reagent. After selecting this window, it is found that H_2SO_4 is not added alone to the mixture, but instead with KBr or some sort of tetraalkylammonium bromide salt. Such conditions would not be immediately obvious without a deep understanding of chemical transformations. It is unfortunate that the reagents are listed alone since such allows conditions such as these to be overlooked. Once again, the computer-based searches are more useful to someone with some foreknowledge of reactions.

Also see Reaction 39 in Chapter 8.

d. The search for 2-bromobutane to butan-2-ol returned four reactions out of four citations. Note that the transformations require reagents other than just hydroxide. If a generic halide (X) is used rather than 2-bromobutane, the search returned 9 reactions out of 17 substances and 13 citations. No reactions for this transformation are reported in Chapter 8.

Performing the 2-bromobutane to 2-butanol search with SciFinder returned only two hits. One reaction used a complicated coded reagent along with water, and the second used water. The searches starting with 2-chlorobutane and 2-iodobutane both returned zero hits.

e. The search for but-2-ene to 2-bromobutane returned six reactions out of four citations. The reaction with HBr under acid conditions was included as a traditional example. The same search with a generic halide (X) rather than 2-bromobutane and a squiggle line to indicate a mixture of (E/Z) isomers returned 18 reactions out of 23 substances and 16 citations.

Searching for the conversion of *trans*-2-butene to 2-bromobutane furnished six hits. Strangely, the traditional reagent HBr was not included in these six hits. Changing the product to the chloride returned 14 hits, and this time, the expected HCl was included as one of the hits.

Also see Reaction 41 in Chapter 8.

f. The search for 2-bromobutane to but-2-ene, using the squiggly bond so that formation of both (E) and (Z) isomers are examined, returned 21 reactions out of 16 citations. The example shown illustrates a classical E2 reaction. The same search with a generic halide (X) rather than 2-bromobutane and a squiggle line to indicate a mixture of E/Z isomers returned 33 reactions out of 38 substances and 29 citations.

When the same reaction is queried on SciFinder, 14 hits are returned. A variety of conditions are returned, including the use of trimethylamine, a relatively weak base. The reaction figures that were returned also clearly show the formation of the (E/Z) isomers in some cases, as well as cases where but-1-ene was formed. When the chloride was the starting material, there were only four hits returned, all using LDA or alkoxide bases.

Also see Reaction 18 in Chapter 8.

g. The search for but-2-ene [using a squiggle line to indicate a mixture of (E/Z) isomers] to but-2-yne returned one reaction out of four substances and one citation, but it is from 1890 and does not provide many details. Changing the search to the reaction of (E/Z) pent-2-ene to give pent-2-yne, returned one reaction out of substances and one citation (from 1877). A new search using but-2-ene to but-2-yne as the basic structure, but adding any group or hydrogen (GH) to carbon rather than methyl (GH− rather than CH_3−) returned 109,747 reactions out of 89,982 substances and 14,707 citations, which is more daunting for a search, but may lead to interesting ideas for new reactions.

With SciFinder, the but-2-ene to but-2-yne search yielded nothing of any use. The only item returned was a paper about selective hydrogenation. However, changing the search to the conversion of pent-2-ene to pent-2-yne furnished one hit, the expected sequential formation of the 2,3-dibromopentane using Br_2 and double elimination with a strong base.

Also see Reaction 19 in Chapter 8.

h. The search for but-2-yne to but-2-ene returned 17 reactions out of 2 substances and 22 citations. Catalytic hydrogenation with different transition metal catalysts are used for the transformation. The generic search using GH rather than methyl returned 246,088 reactions out of 201,205 substances and 0867 citations.

Using SciFinder, a search for the conversion of pent-2-yne into pent-2-ene returned 14 hits. The vast majority of these hits involved hydrogenation to give (2Z)-pentene. Some employ $LiAlH_4$ to furnish the (E)-alkene. None appear to use dissolving metal reduction.

Also see Reaction 17 in Chapter 8.

i. The search for but-2-yne to 2-bromobut-2-ene returned three reactions out of six substances and four citations, including an example for the simple reaction of the alkyne with HBr. The search with a generic halide rather than the bromide returned five reactions out of 10 substances and six citations.

With SciFinder, the same search gave no hits. When chloride was substituted for bromine, the search also returned no hits. If a "substructure" search is executed, almost 30,000 hits are returned, which is unmanageable. Some of the returned hits have additional functional groups in the molecule, while others simply elongate the chain and still others exhibit both characteristics. Conditions include the expected HBr as well as catecholborane/Br_2 mixtures. It is not at all clear why the Reaxys search returned hits while SciFinder failed to return any. It does bring up an important issue however with computer searches. They are limited to their databases and also, more importantly, to the ability of the user to understand and use all of the features. If an error was made entering a reaction into the database, it is likely that the reaction will *never* be found. Likewise, if even a small error is made entering something into the search, it is likely to fail. Looking carefully at the searches performed in this example, these issues are not a problem. Therefore, it is not clear why the search failed. The point of this discussion is that one should utilize all search capabilities of a search engine, as well as any other avenues of research if one computer search fails. Indeed, to truly harness the power of these searches, one often must perform multiple searches.

Also see Reaction 43 in Chapter 8.

j. The search for 2-bromobut-2-ene to but-2-yne returned three reactions out of six substances and three citations, including a simple E2-type reaction. The search using a generic halide rather than the bromide also returned three reactions out of six substances and three citations.

   The elimination of the bromide in 2-bromobut-2-ene to but-2-yne returned no hits on SciFinder, and a search using the iodide likewise returned no hits. The chloride elimination, however, returned three hits. As expected, conditions consistent with an E2 reaction, using $NaNH_2$ as a base are included in the results.

   Also see Reaction 19 in Chapter 8.

k. The search for 2-bromobut-2-ene to butan-2-one returned 0 hits, but 167,781 reactions out of 146,253 substances and 35,555 citations in the expanded search. The search was changed to 2-halobut-2-ene to butan-2-one, using the generic halogen X term rather than the specific Br. This search returned one reaction out of two substances and two citations, illustrating that making the search too specific can cause problems, as well as the fact that using the generic search maybe more useful. However, it is probably prudent to do the specific search first in order to limit the number of examples. In this particular reaction, hydrolysis of the vinyl chloride liberates the ketone. No examples are reported in Chapter 8.

   With SciFinder, searching for specific reactions failed in this case as well. Even a "substructure" search, although it returned an enormous number of hits, did not provide hits in which the vinyl halide was clearly converted to a ketone. When the conversion of (2Z)-chloro-pent-2-ene to pentan-2-one was used for a "substructure" search, many hits were returned. This time, however, at least one was obviously a conversion of a vinyl halide to a ketone. A variety of protic acids or Lewis acids were employed for this purpose.

l. The search for butan-2-one to 2-halobut-2-ene, using the generic search with X given the results from (k), returned two reactions out of seven substances and six citations. One example shows the conversion of butan-2-one to the vinyl chloride by treatment with the halogenating agent, PCl_5. No examples are reported in Chapter 8.

   Searching in SciFinder for the conversion of pentan-2-one to (2Z)-chloropent-2-ene furnished only one hit, using PCl_5, and the stated yield was low (35%) along with 30% 2,2-dichloropentane. Thus, although this reaction is known, one may not expect this reaction to be very useful.

m. The search for butan-2-one to butane returned four reactions out of nine citations. One example shown used hydrogenolysis for the conversion, although both the Wolff-Kishner reduction[2] and the Clemmensen reduction[3] are possible. Note that this search did not return either of those traditional reactions.

With SciFinder, searching for the same reaction gave six hits. Hydrogenation was one choice, but neither the Clemmensen reduction nor the Wolf-Kishner reduction were returned as possibilities. One must question why this is the case. One possible answer may be that the product, butane, is a gas and using butan-2-one as the prototype ketone is a poor choice. At the very least, the conditions to isolate the product for this specific reaction are probably not relevant to the isolation of non-gaseous product in most examples of this transformation. A potentially more relevant search is the conversion of benzophenone to diphenylmethane. This search yields 132 reactions. Of these, the analysis indicates that five hits use HCl (which is consistent with the Clemmensen reduction) and a total of six hits use either N_2H_4 or the hydrated version of hydrazine. Both are consistent with the Wolf-Kishner reduction. An additional 23 hits use hydrogenation. Since diphenylmethane is a solid, the conditions described in any of these hits may be more useful than the conversions to butane. This observation suggests that not only insight into the chemical reactions will be helpful when performing searches, but also knowledge of the physical states of reagents and products.

Also see Reaction 16 in Chapter 8.

n. The search for butan-2-ol to butane returned five reactions out of four citations. Note the rather specialized reaction conditions required for this particular example. One can also imagine conversion of the alcohol to an alkene by dehydration, followed by catalytic hydrogenation.

When the conversion of butan-2-ol to butane is searched using SciFinder, 26 hits were returned. The conditions appear to employ

[2] (a) Kishner, N. J. *Russ. Phys. Chem. Soc.* **1911**, *43*, 582;(b) Wolff, L. *Annalen,* **1912**, *394*, 86; (c) Todd, D. *Org. React.* **1948**, *4*, 378; (d) The Merck Index, 14th ed., Merck & Co., Inc., Whitehouse Station, NJ, **2006**, p ONR-103; (e) Mundy, B.P.; Ellerd, M.G.; Favaloro Jr., F.G. *Name Reactions and Reagents in Organic Synthesis,* 2nd ed., Wiley-Interscience, NJ, **2005**, pp. 704-705.

[3] (a) Clemmensen, E. *Berichte* **1913**, *46*, 1837;(b) Vedejs, E. *Org React.* **1975**, *22*, 401; (c) The Merck Index, 14th ed., Merck & Co., Inc., Whitehouse Station, NJ, **2006**, p ONR-18; (d) Mundy, B.P.; Ellerd, M.G.; Favaloro Jr., F.G. *Name Reactions and Reagents in Organic Synthesis,* 2nd ed., Wiley-Interscience, NJ, **2005**, pp. 160-161.

very high pressures and temperatures. Using the "analyze" function shows H_2 to be a reagent. The two associated hits *do not* involve elimination to the alkene, followed by hydrogenation. As in (m), however, butane is a product so the isolation procedure is likely to be atypical. Expanding the search to use the "substructure" tool gave nearly 10 million hits, useless to an almost comical degree. Changing the target to search the conversion of cyclohexylmethanol to methylcyclohexane returned three hits. Of these three hits, one appears to deliberately make the methylcyclohexane, while the others appear to be simply *a* product of a reaction where cyclohexylmethanol was the starting material. One of the yields is reported as 1% and the other as 5%. Thus before selecting a reference, the yields reported for the desired target should be considered. If the yields are not given, it is probably wise to obtain additional references and perhaps even discard references/hits that do not report a yield.

Also see Reactions 15 and 16 in Chapter 8.

o. The search for 2-halobutane to butane, used the generic X group for the halide, and returned 39 reactions out of 25 citations. The hydrogenolysis of an iodide moiety was shown with several reagents.

When searching for 2-bromobutane to butane using SciFinder, six hits were returned but only two appear to directly accomplish the desired transformation. Similar to the previous cases, searching for a more sophisticated structure reveals other, perhaps more useful reactions. For example, a search for 1-bromo-1-phenylethane to ethyl benzene returned reactions that used $NaBH_3CN$ and other reducing agents.

Also see Reaction 15 in Chapter 8.

p. The search for butane to 2-halobutane used the halobutane molecule with the generic X group, returned 14 reactions out of 27 citations. The radical bromination of an alkene using NBS was one example.

When the same search was done with SciFinder, six reactions were returned. The reactions all appear to involve radical halogenation, as expected. Keep in mind, however, that although the conditions ought to be the same with gaseous reagents, one would likely search for other reactions if an experimental procedure is needed if the target were not also a gas.

Also see Reaction 40 in Chapter 8.

q. The search for the reaction of (*E*/*Z*)-but-2-ene to butane returned 55 reactions out of 60 substances and 67 citations. Catalytic hydrogenation is the traditional method to accomplish this transformation.

The same search using SciFinder returned 72 reactions. As expected, a large number of the reactions directly involve catalytic hydrogenation. Also see Reaction 14 in Chapter 8.

r. The search for but-2-yne to butane returned five reactions out of 10 citations. One example illustrated the use of catalytic hydrogenation for this conversion.

As with the alkene in (q), this transformation can be accomplished with hydrogenation. SciFinder returns six hits for this transformation. In this search, several less traditional reagents were returned along with hydrogenation.

Also see Reactions 14 and 17 in Chapter 8.

s. The search for (E/Z)-2-halobut-2-ene to butane, using the generic X group rather than the Br, and returned 0 hits, and the automatic expanded search returned about 2,770,440 reactions out of 1,872,492 substances and 224,757 citations. The search was modified to search the reaction 1-halocyclohexene to cyclohexane. This search returned three reactions out of five substances and two citations, which includes a hydrogenolysis example.

Also see Reaction 14 in Chapter 8.

t. The search for butan-2-one to 2-methylbut-1-ene returned one reaction with one citation, but this example involved thermolysis and is not general. Therefore, the search was modified to use the reaction of cyclohexanone to methylenecyclohexane, and this search returned 24 reactions out of 24 substances and 24 citations. No examples are shown for this transformation in Chapter 8.

The 2-butane to 2-methylbut-2-ene search on SciFinder returned zero reactions. When the cyclohexanone reaction was searched, 79 hits we returned including Peterson olefination and Wittig reactions. Curiously, none of the hits appeared to be a McMurry coupling, which converts two carbonyls into alkenes using Ti salts.[4]

u. The search of butan-2-one to 2-ethyl-2-methyloxirane returned zero hits, and the expanded search returned 36,712 reactions out of 6447 citations. While some of the hits are related to the initial search, the reaction was modified to search the conversion of cyclohexanone to exocyclic spirocyclic epoxide. This search returned 21 reactions out of 24 citations, including one example shown in which the ketone reacts with a sulfur ylid to give the epoxide. No examples are shown in Chapter 8.

[4] See McMurry, J.E.; Rico, J.G. *Tetrahedron Lett.* **1989**, *30*, 1169.

v. The search of butan-2-one to but-2-ene to 2,3-dimethyloxirane returned four reactions out of four citations. One example that was returned involved epoxidation of the alkene using peroxy acids.

    With SciFinder, searching for the direct conversion of cyclohexene to cyclohexene oxide provided 2603 hits. A large variety of reagents were returned, although the analysis indicates just over half of the returned reactions employed H_2O_2 or t-BuOOH.

    Also see Reaction 16 in Chapter 8.

w. The search for bu-2-ene to butan-2-ol returned 14 reactions out of 27 citations. One example that was returned illustrated hydration of the alkene to give the alcohol. Remember that hydroboration of an alkene usually gives the less substituted alcohol, whereas hydration generally gives the more substituted alcohol.

    Searching the same reaction using SciFinder returned 64 hits. Typical acid—base reactions were the most common examples. In the "analysis" window, some reagents, including a chiral dialkylborane, are also returned that provide stereoselective reaction conditions. As described for the Reaxys search, hydroboration furnishes the less substituted alcohol from an unsymmetrical alkene. In cases where the alkene carbons are equally substituted, no such selectivity is possible in the absence of steric or chiral directing groups.

    Also see Reaction 04 in Chapter 8.

x. Based on the results returned for (t), the search using the reaction of methylenecyclohexane to cyclohexanone returned 10 reactions out of 9 citations. One example illustrated the ozonolysis of the alkene moiety to generate the ketone.

    The same search with SciFinder gave 13 hits. Although other conditions were returned as well, ozonolysis was one of them, as expected.

    Also see Reactions 10 and 12 in Chapter 8.

One of the benefits of this complicated search profile for many reactions is that it makes clear some transformations are rather straightforward whereas others are rare or more commonly require multiple steps. Such information is important for any retrosynthetic plan. Another important issue shown by these searches is that the choice of reaction to be searched is essential. In some cases, initial results showed no hits, or returned reactions that are impractical and not very general. It is a graphic illustration that chemical intuition is essential, requiring proper training via coursework or practical training. Finally, Figure 5.1 shows that several functional groups can be converted to the corresponding carboxylic acid derivative, allowing acyl substitution reactions involving acid chlorides, esters, amides or anhydrides. A useful exercise would involve using Reaxys to probe the efficacy of these transformations.

It is important to emphasize that Figure 5.1 is *not* intended as a device to memorize specific reactions, but rather to introduce the idea of using various functional groups in a synthesis by recognizing their synthetic relationships. A good textbook or the use of Reaxys or SciFinder can supply a variety of reagents/conditions, including more modern techniques for each transformation. In complex systems, simple reagents may not work very well, and finding reasonable alternatives is usually important.

CHAPTER 6

Stereochemistry

An important characteristic in a target is the presence of one or more stereogenic centers. Target **1** in Figure 6.1 has one stereogenic center, although *the racemic molecule is shown as the target*. As noted previously, a search of each specific enantiomer may not yield useful results, or indeed any results. Using computer-based searches, it is reasonable to search both enantiomers, as well as the racemate, but it is probably more productive to first search the racemic compounds. Using **1** as an example, a reasonable disconnection invokes the aldol condensation reaction (Reaction 80 in Chapter 8), with disconnect products **2** and **3**. Indeed, both **2** and **3** are commercially available. For the purposes of this example, search racemic **1**, followed by (S)-**1** and then (R)-**1**.

A Reaxys search of racemic **1** returned two reactions in three citations, shown in Figure 6.2. The results shown clearly indicate that an aldol condensation (Reaction 80 in Chapter 8) is used to prepare **1**, and also that the product is racemic in the second example. One of the reactions shown for the first example is listed as an enantioselective reaction, so one expects a mixture that favors one enantiomer.

A Reaxys search using (3S)-**1** gave the results shown in Figure 6.2, with the report by Trost et al. and that by Li et al. In both cases, the enantiomer is formed as part of a mixture, so it was reported to be an enantioselective and not an enantiospecific reaction. This result focused on the (3S)-enantiomer, formed as part of racemic mixture, which is no help if an asymmetric synthesis is the target. A similar search for the (3R)-**1** gave the same result.

This Reaxys search suggested that no synthesis of enantiopure (3R) or (3S)-**1** is available. Indeed, if one enantiomer is searched, it is possible there

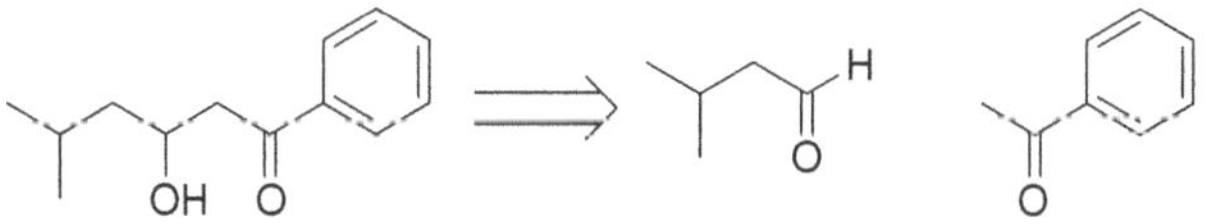

Figure 6.1 Disconnection of racemic **1**.

Hybrid Retrosynthesis.
DOI: http://dx.doi.org/10.1016/B978-0-12-411498-2.00006-1

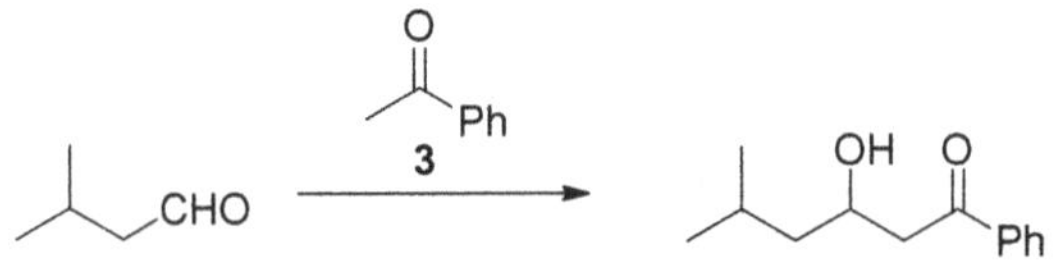

Figure 6.2 Reaxys returns for a synthesis of **1** from **2**.

1. With molecular sieve 4 Å; triphenylphosphine sulfide; 2,6-bis(2R)-2-(2,2-diphenylcarbonol)azolidine)-p-cresol diethylzinc in THF; hexane. R = –5°C, 48 h. This compound not separated from byproducts.
Trost, Barry M.; Ito, Hisanaka Journal of the American Chemical Society, 2000, vol. 122, #48, p. 12003–12004.

With C56H52N2O4; diethylzinc; triethylamine in hexane; N,N-dimethylformamide. T = 0°; 120 h; Inert atmosphere. Molecular sieve; optical yield given as percent ee enantioselective reaction.
Li, Hong; Da, Chao-Shan; Xiao, Yu-Hua; Li, Xiao; Su, Ya-Ning Journal of Organic Chemistry, 2008, vol. 73, #18, p. 7398–7401,

2. Stage #1: Acetophenone With lithium diisopropylamide in THF. T = –78°C; 0.5h; Inert atmpshere.   Stage #2: isovaleraldehdye in THF. T = –78°C; 0.5 h; Inert atmosphere. Provencher, Bria A.; Bartelson, Keith J. Liu, Yan; Foxman, Bruce M.; Deng,Li Angewandte Chemie, Internatnal Edition, 2011, vol. 50, #45, p. 10565–10569; Angewandte Chemie, 2011, vol. 123, #45, p. 10753–10757.

will be no hit at all for a given molecule. The synthetic chemist must search other methods that generate the target with good enantioselectivity and then apply them to the problem at hand.

Another example illustrates yet another point about searches of this type. It is known from other work by the authors that **4** has been prepared with good enantioselectivity by Northrup and MacMillan.[1] A Reaxys search was performed using **4** with the specific stereochemistry shown, and this search returned one reaction out of three substances and four citations. As shown, this search showed an asymmetric synthesis of **4**, using L-proline as a chiral additive, with four different references using somewhat different reaction conditions. Interestingly, the citation of MacMillan's work was not displayed in the screen shot. However, clicking on "full text" under the patent reveals that this is the work by MacMillan (Figure 6.3).

This latter observation allows one to probe another feature of Reaxys, which has not been used previously in this book, but may be quite useful. If one clicks "find similar reactions," several new items are displayed, including the 2002 *J. Am. Chem. Soc.* reference by MacMillan. It is not clear why this MacMillan article was not displayed the first time, but the lesson here is twofold. First, it is worthwhile to search the specific stereoisomer of interest, although it may be necessary to search the racemate as well. Second, it may be useful to explore the links to displayed articles for details, full text, similar reactions or even citing

[1] Northrup, A.B.; MacMillan, D.W.C. *J. Am. Chem. Soc.* **2002**, *124*, 6798.

4

1. 88% yield. With L-proline in NAN-dimethyl-formamide T=4°C; 1.6 h; Product distribution / selectivity; CALIFORNIA INSTITUTE OF TECHNOLOGY Patent: WO2003/89396 A1, 2003; Location in patent: Page/Page column 21-22; 26-27;

2. 72% yield. With 1-n-butyl-3-methylimidazolium hexafluorophosphate; L-proline in N,N-dimethyl-formamide T=4°C; Cordova, Armando Tetrahedron Letters, 2004, vol. 45, # 20 p. 3949–3952

3. With L-proline in N,N-dimethyl-formamide T=4°C; 16 h; Casas, Jesus; Engqvist, Magnus; Ibrahem, Ismail; Kaynak, Betul; Cordova, Armando Angewandte Chemie - International Edition, 2005, vol. 44, # 9 p. 1343–1345. Cordova, Armando; Ibrahem, Ismail; Casas, Jesus; Sunden, Henrik; Engqvist, Magnus; Reyes, Efraim Chemistry - A European Journal, 2005, vol. 11, # 16 p. 4772–4784

Figure 6.3 Reaxys returns for a synthesis of **4**.

articles for the displayed references. In other words, use all tools available to provide as much information as possible.

From the standpoint of a more general discussion of retrosynthesis, there is another issue that should be addressed. There are usually many targets that could be disconnected in more than one acceptable manner to give different retrosynthesis, and therefore different syntheses. Disconnection on either side of the carbonyl in **1** is reasonable, for example, as well as either side of the CH−OH moiety. The disconnection in Figure 6.1 to give **2** and **3** is preferred because it points to an aldol condensation (see Reaction 80 in Chapter 8), but there is another reason for choosing this disconnection. *Note that one bond α to the C=O and two bonds β to C=O are connected to the stereogenic center adjacent to the carbonyl.*

> *When possible, disconnect a bond that is connected to a stereogenic center.*

There is a good rationale for this statement. Formation of the stereogenic center during a reaction offers the potential to control the stereochemistry of that center. Disconnection of a bond that is not attached to a stereogenic center simply means that the stereogenic center must be made in another reaction. The stereogenic center in the fragment must be dealt with sooner or later. Sooner is better than later in a synthesis. Note that *the synthesis will generate racemic products and the focus is not on controlling absolute stereochemistry. Therefore, diastereoselectivity is an important issue but not enantioselectivity.*

Although other disconnections may be perfectly reasonable, if the disconnection generates a compound that has a stereogenic center there is potentially a problem. If that compound is not available in chiral form, then it must be prepared by a reaction that gives an enantiopure product, which may be difficult. Forming the stereogenic center directly in a reaction is usually preferable because it may be possible to control the stereoselectivity of that reaction. In other words, *when there is a choice, choose a disconnection that generates the stereogenic center directly rather than a disconnection that leaves a stereogenic center on the disconnect product.*

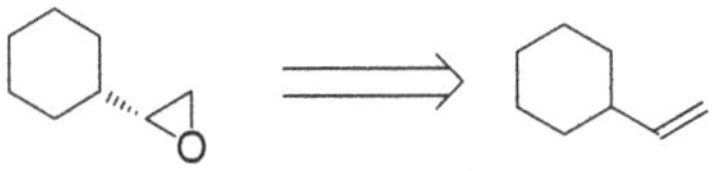

The targets chosen in this section focused on stereochemistry, and problems relating to exact match structure searches were highlighted. Reaxys is certainly capable to providing synthetic routes to chiral targets, as is SciFinder. A simple example is the chiral epoxide **5**. A structure specific search using Reaxys returned 17 reactions out of 20 substances, with 23 citations. One showed the enantioselective synthesis of **5** from **6**, by heating **6** with the salen catalyst **7**, Ti(OiPr)$_4$, and hydrogen peroxide in dichloromethane/dichloroethane.[2] Rather than give many rather simple examples, Chapter 7 is provided to show that molecules with important stereochemistry can be searched and useful and critical hits are returned for these compounds.

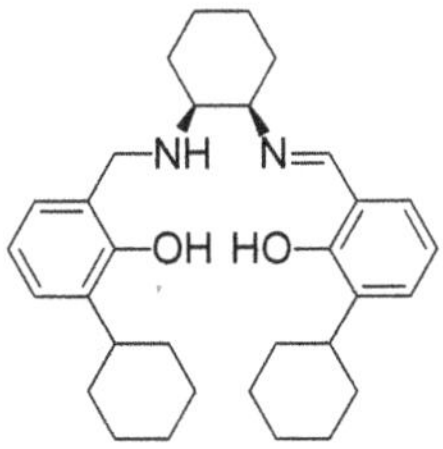

---

[2] Berkessel, A.; Günther, T.; Wang, Q.; Neudörfl, J.-M. *Angew. Chem. Int. Ed.* **2013**, *52*, 8467.

CHAPTER 7

Molecules of Greater Complexity

Contents

7.A CONVERGENT SYNTHESES

The synthesis of complex molecules requires an in-depth knowledge of chemical reactions, as well as an understanding of synthetic strategy for various types of structural motifs. That is, it also requires an understanding of the order in which the chemical transformations must be done. An entire course is usually required just to introduce the fundamentals of synthesis theory and a lifetime of experience to master it. One aspect of a planned synthesis concerns strategy; will a single starting material be transformed to the target in a linear manner, or will the target be broken into several key fragments, each one synthesized and then assembled to complete the synthesis? In other words, will be synthesis be consecutive (i.e., linear) or convergent?

Compound **1** in Figure 7.1 illustrates a slightly more complex target, and one that offers a possibility for different synthetic strategies. This brief analysis will focus on the different strategies rather than specific syntheses. Disconnection of bond *b* leads to disconnect products **2** and **3**. There is no attempt here to translate these products into real molecules. Again, the intent is to illustrate a strategy, not specific reactions. Compare this disconnection with the disconnection of bond *a*, which gives **4** and **5**. Disconnection of bond *b* cleaves **1** into two fragments that are close in mass (C_6O for **2** and C_9 for **3**), but disconnection of bond *a* gives a small fragment (**4**) and a large fragment, **5**. In effect, disconnection of bond *b* leads to two equivalent fragments. Why is the relative size of the disconnection fragments important? The first disconnection gives fragments that are combined in the last chemical step of the synthesis. For disconnection of bond *b*, coupling the synthetic

Hybrid Retrosynthesis.
DOI: http://dx.doi.org/10.1016/B978-0-12-411498-2.00007-3

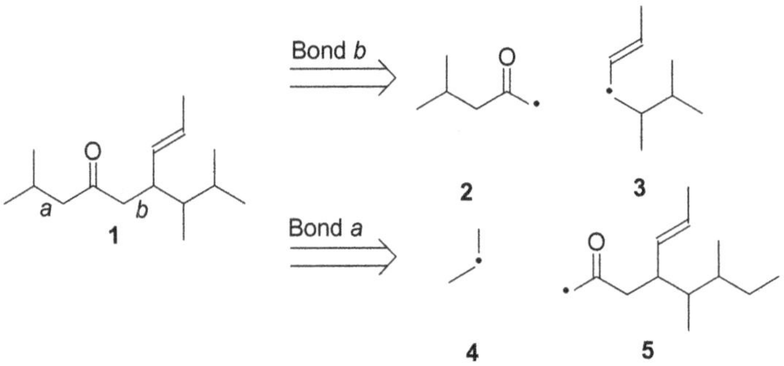

Figure 7.1 Disconnection of bonds *a* and *b* in **1**.

equivalents for **2** and **3** will give **1**. Likewise, coupling the synthetic equivalents for **4** and **5** will give **1**. Assume that **2**, **3**, and **5** all require further disconnection. It is likely that the disconnection and synthesis of **2** or **3** will require fewer steps than disconnection and synthesis of **5**. Therefore, a synthesis of **1** via combining fragments **2** and **3** is likely to be shorter, and more efficient. Another way to make this statement is to say that *the goal is to correlate a disconnection with simplification of the target.* Greater simplification of the target in a disconnection should lead to a more efficient synthesis. Based on this idea, the disconnection of bond *b* provides more simplification of **1** than the disconnection of bond *a*, whereas disconnection of bond *a* only trims a small piece from **1**.

Disconnection of target **1** to **2** and **3** gives more simplification, and it is stated that this leads to a more efficient synthesis. Why is this more efficient? An answer to this question requires two definitions: convergent synthesis and consecutive (i.e., linear) synthesis. One synthetic strategy is to work backwards linearly from the target to a starting material. A synthesis based on a linear pathway is called a *consecutive synthesis*. The consecutive approach is used in previous examples in this chapter for all disconnections. Alternatively, the synthetic strategy may disconnect to several fragments, each with its own starting material. The resulting disconnect fragments are combined in a non-linear fashion to prepare the target, and a synthesis based on this branching pathway is called a *convergent synthesis*. In a convergent synthesis, the several pieces of a molecule are synthesized individually, and the final target is sequentially assembled from the pieces. For **1**, fragments **2** and **3** may be synthesized individually and then combined to give **1**. This is a simple example of a convergent synthesis.

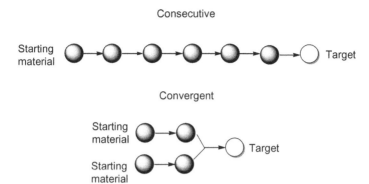

Figure 7.2 A comparison of a consecutive versus a convergent synthesis.

Figure 7.2 illustrates the difference between a *consecutive* and a *convergent* synthesis. In the consecutive synthesis, a single starting material is converted stepwise to the target in six steps. In the convergent synthesis, there are two starting materials and each is converted to a key fragment in one step. The two key fragments are combined in one step to give the target. This disconnection constitutes a shorter and more efficient synthesis, and it should yield a greater amount of product. It is suggested that one use a convergent strategy when possible, and a convergent strategy requires disconnections that will provide the most simplification. In the case of **1**, disconnection of bond *b* is preferred. It is likely that screening reactions for various disconnections will provide information about consecutive versus convergent strategies, or different routes within each manifold.

7.B PEROVSKONE

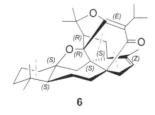

6

The techniques discussed in previous chapters are designed to examine relatively simple synthetic targets. Indeed, the examples presented are mostly based on reactions and problems that arise in an undergraduate organic chemistry course. Complex targets such as natural products usually have >10—15 carbon atoms, multiple functional groups, multiple stereogenic centers, and structural complexity that are not necessarily solved by a simple approach. Nonetheless, fundamental approaches for bond disconnection and choices of library reactions can be applied to a given problem. However, a much greater understanding of reactions is required, including cutting-edge reaction techniques that are only obtained from a thorough understanding of the literature, and the development of excellent laboratory skills and instincts.

Perovskone (6) is clearly a significantly more complex target than any discussed to this point. Is it possible to use a computer search to address issues related to such a complex target? G. Majetich reported a total synthesis of perovskone, and it is found in the *Journal of the American Chemical Society*.[1] It is a known synthesis, and a search of this structure without the stereochemistry using Reaxys returned 26 reactions out of 25 substances and 2 citations, including reference 1 and a second paper by Majetich.[2] The authors of this work knew the synthesis of perovskone, and there was a high expectation that it would be in the database. A confession of sorts is in order. While putting structure 6 into Reaxys, a methyl group was omitted from the structure, and the exact structure search returned zero hits. This *mea culpa* is added to emphasize that it is important to check to be certain that the exact structure has been entered, particularly with very complex molecules. If the structure of such a complex molecule has not been synthesized, it is likely that a search will return zero hits, and the use of the hybrid-retrosynthesis approach is probably essential.

Figure 7.3 shows the first disconnection taken from the retrosynthetic analysis of 6 in the reported synthesis by Majetich.[1,2] Although the search with Reaxys using 6 as a target returned the actual synthesis, the hybrid retrosynthesis approach can be tested. If another search is done using disconnect compound 7 from Figure 7.3, 38 reactions out of 2 citations are returned, but once again both cite the work of Majetich. The point of this exercise is that simplifying the structure and using the hybrid approach introduced in previous chapters also provided a solution to the problem. In this case, a search of an intermediate taken from the

[1] Majetich, G.; Zhang, Y. *J. Am. Chem. Soc.* **1994**, *116*, 4979.
[2] Majetich, G.; Zhang, Y.; Tian, X.; Britton, J.E.; Li, Y.; Phillips, R. *Tetrahedron* **2011**, *67*, 10129.

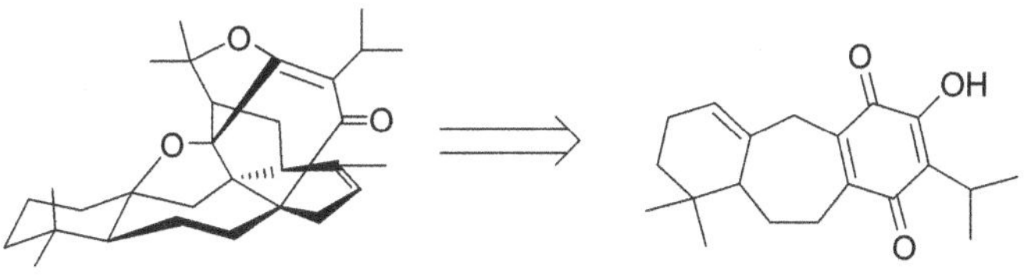

Figure 7.3 First disconnection from the Majetich synthesis of **6**.

retrosynthesis allowed one to not only find the synthesis of **6**, but also to examine the synthesis of a key intermediate. Further searches on intermediates might lead to alternative syntheses for some of them, and possibly an idea for a different synthesis. As in many cases, however, chemical intuition and a flexible search strategy is necessary for success in a synthesis of such a complex target.

Searching for the structure of perovskone using SciFinder led to several hits that were immediately returned. If the user clicks on one of the two structures (one showed absolute stereochemistry, the other, relative), a very vague synthesis is displayed. Although difficult to read in the displayed format, a link directly to the references is provided, as well as an indication of which of the starting materials employed in the synthesis were commercially available. For all entities, hovering over the structure brings up an option window by clicking the $\gg$ icon. One option is "get commercial sources." Clicking this option brings the user to a page that lists all of the commercial sources for the substance. For some suppliers, there is even an "order from source" link on this page, along with some pricing. For cases where the molecule is not commercially available, or cases where it is prohibitively expensive, the button "synthesize this" can be selected, instead. *Of course, this approach will only work if the molecule has been synthesized,* which the vast majority of compounds returning a hit have been. There are cases where the molecule may not have been synthesized, but hits are returned if the target is a newly discovered natural product, the structure of which has been reported but its synthesis still eludes chemists.

One significantly powerful utility offered by SciFinder is searching for not just structures or reactions but also for topics. When perovskone was searched for as a term, 11 hits were returned. The hits include not just multiple syntheses but also the original isolation paper. For a novel or otherwise unnamed structure, of course, such a search would return zero hits.

7.C AMPHOTERICIN B

8

Amphotericin B (**8**) is an antifungal agent produced by *Streptomyces nodus*. Amphotericin B has a very complex structure, but the total synthesis has been reported by a number of groups. It is a macrolactone with seven contiguous *trans* double bonds; 14 stereogenic (chiral) centers; and there is a cyclic hemi-acetal moiety. With the complexity of **8** in mind, the option to search by the topic *Amphotericin B* in SciFinder is particularly attractive. This search returned over 31,000 hits, which for all purposes is unmanageable. When the search term for a topic search was changed to *total synthesis of amphotericin B*, only 31 references were returned. Any one of these should provide references that report the synthesis of this complex molecule. When the search parameter was changed to a structure search, using **8** as the target, several hits were returned. When "synthesize this" was chosen, a large number of reactions were returned. The images were extremely difficult to read, due to the fact that the structure of a very large molecule was compressed to fit on the page. This difficulty is at least due, in part, to the layout of the data. Even if one "zooms" in using a browser, the difficulty in reading the structure persists, but one can access one or more of the original papers for a better reading. Reverting back to a text-based search, individual papers can be easily selected and different total syntheses can be examined. Such references could also be found for the preparation of any of the intermediates if one chooses though it is unclear what incentive there would be since all of this information would be contained in the reference(s) chosen.

When comparing the syntheses reported by Nicolaou[3] and by McGarvey[4] for amphotericin B, both syntheses identified **9** as an

---

[3] Nicolaou, K.C.; Oglivie, W.W. *Chemtracts-Organic Chemistry* **1990**, *3*, 327–349.
[4] McGarvey, G. J.; Mathys, J. A.; Wilson, K. J. *J. Org. Chem.* **1996**, *61*, 5704–5705.

attractive intermediate. This intermediate could be formed via a tandem esterification and Horner-Wadsworth-Emmons reaction[5] from the hydroxy aldehyde **10** and the acid phosphonate ester **11**. Thus, finding common intermediates is possible when taking a hybrid approach to the synthesis of more complex molecules. Each of the references will provide unique syntheses to their targets and both the synthetic strategy and synthetic steps can be gleaned from the references.

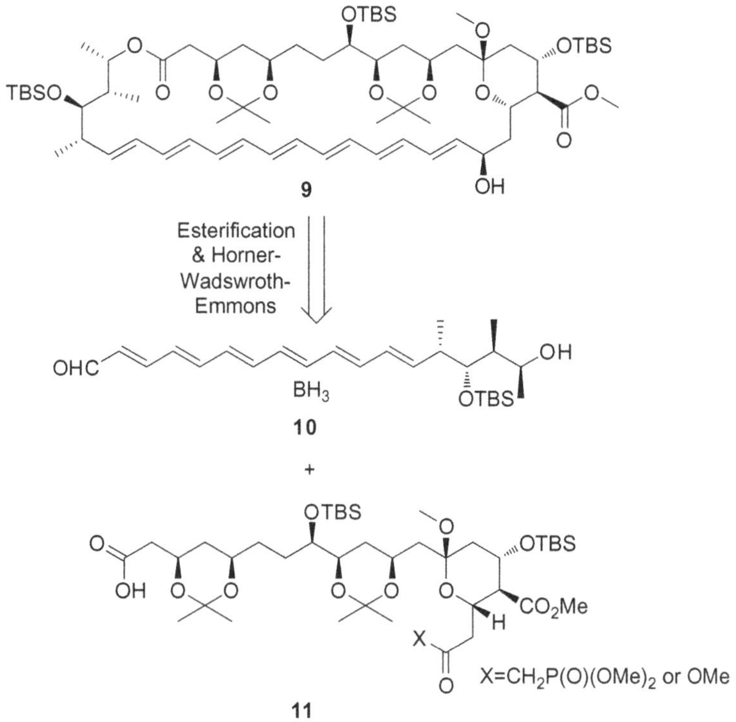

9

Esterification
& Horner-
Wadswroth-
Emmons

BH_3

10

+

$X=CH_2P(O)(OMe)_2$ or OMe

11

7.D CARBOVIR

With the struggles associated with viewing the structures for amphotericin B in SciFinder in mind, a smaller, though no less important target, carbovir (**12**), was examined. Carbovir is a potent anti-viral compound deployed in the fight against HIV. Its mechanism of action is that it acts as a nucleobase reverse transcriptase inhibitor, which means that **12** disrupts

[5] (a) Boutagy, J.; Thomas, R. *Chem. Rev.* **1974**, 74, 87; (b) Mundy, B.P.; Ellerd, M.G.; Favaloro Jr., F.G. *Name Reactions and Reagents in Organic Synthesis,* 2nd ed., Wiley-Interscience, NJ, **2005**, pp. 334−335.

the synthesis of the HIV DNA strand. The simpler structure of **12** in this search led to results using SciFinder that were far easier to read and to navigate than those obtained for amphotericin B (**8**).

12   **13**

If one chooses the same "synthesize this" button available on SciFinder as discussed in previous sections, a number of syntheses are returned. The user then has the option to export the search as a .pdf file. At this point, another option will exist that will allow the user to include details. Exercising this option allows for easy-to-read figures and reference information and also provides detailed experimental procedures and spectroscopic information for many of the syntheses. In those instances where the detailed procedures are not provided, reaction conditions are available along with the reference. The layout of the information is very easy to navigate, with the reference information in a shaded box and copyright information for the search separating one synthesis from the next. One of the common elements of syntheses of **12** is the use of the 2-amino-6-chloropurine **13**. Although synthetic modification to the cyclopentene core of **12** varies with different syntheses, the use of **13** as a precursor to the guanine unit in carbovir is extremely common. It turns out that the use of **13** as a guanine precursor is not exclusive to carbovir. Alkylation of guanine at this N^9 (see **13**) is extremely difficult, even with the use of protecting groups. Using **13** instead greatly increases the yield of alkylation at this position, even without a protecting group on the free amine. The use of aqueous base and heat converts the chloride into a hydroxyl group, which is the tautomer for the amide in **12**.

Another common element in the syntheses of this important pharmaceutical is that the 2-amino-6-chloropurine is introduced into the molecule late in the synthetic sequence. One reason for this fact is due to the enormous increase in polarity and potential water solubility of the product by the incorporation of so many nitrogen atoms into the structure. This complicates the isolation and purification of each intermediate after this step. This observation is not immediately clear if one is not familiar with the properties of the purine and pyrimidine bases. In addition, the amide moiety is subject to tautomerization with the free amine,[6] and there are multiple nucleophiles in this system, meaning they may interfere with earlier chemistry if incorporated into the molecule too early in the synthetic sequence. It is for these reasons that one must be careful of the order reactions are planned, not just which reactions are planned.

---

[6] Smith, M.B., March's Advanced Organic Chemistry, 7th Ed, John Wiley & Sons, NJ, p. 91.

CHAPTER 8

Common Fundamental Reactions in Organic Chemistry

Contents

Hybrid Retrosynthesis.
DOI: http://dx.doi.org/10.1016/B978-0-12-411498-2.00008-5
77

Representative experimental conditions are shown in the following two sections. These procedures provide a link between "paper chemistry" and the practical execution of those reactions in the laboratory. The procedures are not intended as a laboratory manual to do the experiments, but rather a tool to evaluate the efficacy of a transformation and a guide to find appropriate reactions. It has a secondary intent of providing some insight into how difficult a chemical reaction may be, which can never be seen from a simple reaction figure. The goal is to begin the process of evaluating different reagents and reactions that may accomplish the same transformation in a synthesis. Such an evaluation is important for choosing a reagent or, perhaps as importantly, thinking about alternative reagents should the original choice prove unsuitable.

Since the experimental details are not intended as a laboratory manual, the conditions are not always quoted verbatim from the primary literature, and many procedures have been modified with respect to language or represent general procedures rather than a specific procedure. As a result, no attempt should be made to repeat the procedures herein, or similar ones, unless the original paper is consulted. Any other course of action is irresponsible, unsafe, and possibly dangerous. Note that experimental protocols must often be harvested from the supplemental information of the paper referenced herein.

8.A FUNCTIONAL GROUP EXCHANGE REACTIONS

Acetals and Ketals

1 From Aldehydes or Ketones

Aldehydes and ketones react with alcohols to form acetals or ketals.

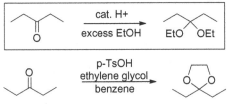

J. Org. Chem. **2008**, *73, 5549–5557.*

Ethylene glycol (0.114 mol) and a few crystals of *p*-toluenesulfonic acid were added to a solution of pentan-3-one (0.0568 mol) in benzene

(50 mL), under argon. The resulting mixture was refluxed using a Dean-Stark apparatus. After a reaction time of 4 h, benzene was distilled at atmospheric pressure. The pressure was reduced and the product was isolated. Note that benzene is a cancer suspect agent.

Acid Chlorides

2 From Carboxylic Acids

Carboxylic acids react with thionyl chloride and other chlorinating agents to give an acid chloride.

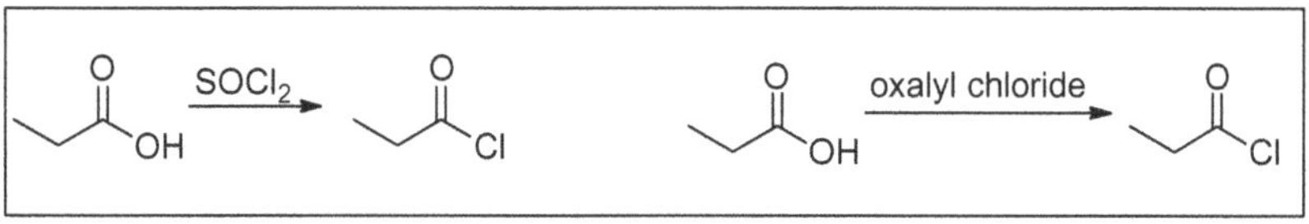

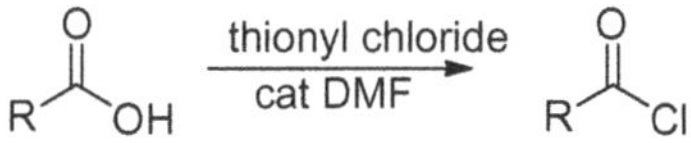

Eur. J. Med. Chem. **2013**, 70, 548–557.

To previously distilled thionyl chloride (5 mL) were added of the carboxylic acid (3 mmol) and immediately 3 drops of DMF. The reaction mixture was heated at 70°C for approximately 90 min, forming a yellowish translucent solution. Excess thionyl chloride was removed from the mixture with a rotary evaporator connected to a potassium hydroxide trap. Products obtained were pale yellow oils or solids. Acyl chlorides were employed directly for the following reactions without further purification.

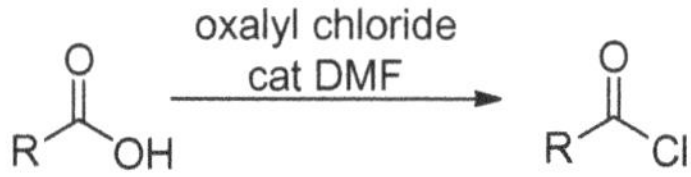

J. Am. Chem. Soc. **2013**, 135, 7442–7445.

A flask was charged with the appropriate carboxylic acid (1 equiv) and dichloromethane. Two drops of DMF and oxalyl chloride (1.2 equiv) were added dropwise. The solution was stirred at 23°C for 3 h and then concentrated *in vacuo*. The crude acid chloride was used without further purification.

Alcohols

3 From Acid Derivatives (acid chlorides, anhydrides, esters)

Acid chlorides, anhydrides, and esters are reduced to alcohols with $LiAlH_4$, H_2 and a catalyst. Conjugated esters undergo a mixture of 1,2- and 1,4-reduction with $LiAlH_4$ or $NaBH_4$.

J. Am. Chem. Soc. **1951**, 73, *107–109.*

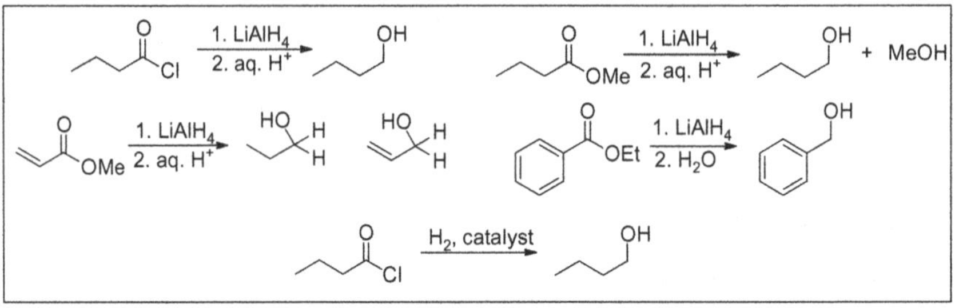

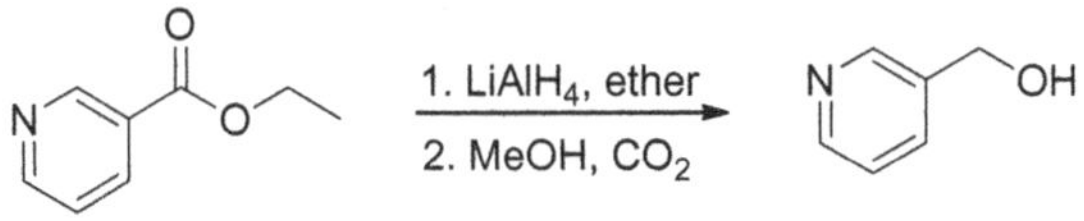

A solution of 14.8 g (0.4 mol) of lithium aluminum hydride in 300 mL of anhydrous ether was placed in a flask provided with a dropping funnel, reflux condenser and stirrer. The solution was stirred, and during a period of one-half hour a solution of 30.2 g (0.2 mol) of ethyl nicotinate in 200 mL of dry ether was added. To the reaction mixture was added dropwise with stirring 50 mL of water to destroy excess lithium aluminum hydride. The mixture was filtered. The solid was suspended in 300 mL of methanol; this mixture was saturated with carbon dioxide, heated to boiling and filtered. Again the solid was extracted with 300 mL of hot methanol. The combined ether and methanol filtrates were evacuated *in vacuo*. The residual liquid was taken up in ether, the solution dried with anhydrous potassium carbonate and distilled to yield 17 g (82%) of 8-hydroxymethylpyridine.

4 From Alkenes

Alkenes react with weak acids such as water and alcohols using a catalytic amount of a strong acid; with mercuric compounds and water to give an alcohol after reduction with $NaBH_4$; with boranes to give alcohols, after treatment with $NaOH/H_2O_2$.

J. Am. Chem. Soc. **1967**, 89, *1522–1524.*

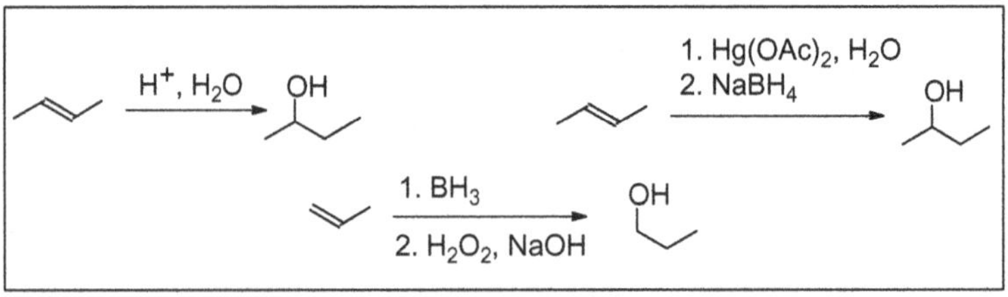

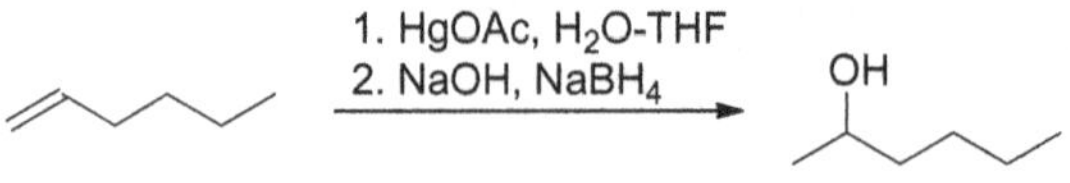

In a small flask, fitted with a magnetic stirrer, is placed mercuric acetate (10 mmol). To this flask is added water (10 mL), followed by THF (10 mL). Then 1-hexene (10 mmol) is added. The reaction mixture is stirred for 10 min (at room temperature, approximately 25°C) to complete the oxymercuration stage. Then 3 M sodium hydroxide (10 mL) is added, followed 0.5 M sodium borohydride in 3.0 M sodium hydroxide (10 mL). Reduction of the oxymercurial is almost instantaneous. The mercury is allowed to settle. Sodium chloride is added to saturate the water layer. The upper layer of THF is separated, and it contains essentially a quantitative yield of hexan-2-ol.

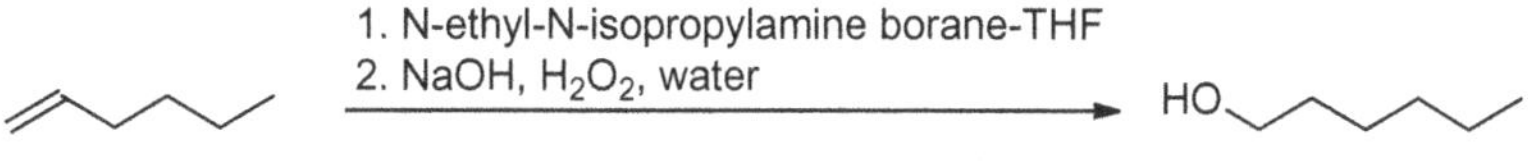

1. N-ethyl-N-isopropylamine borane-THF
2. NaOH, H$_2$O$_2$, water

J. Org. Chem. **1998**, 63, *5154–5163.*

An oven-dried hydroboration flask was cooled to 0°C under a stream of nitrogen. In the flask was placed N-ethyl-N-isopropyl aniline borane (5 mmol) in freshly distilled THF (7.5 mL) and undecane (7.5 mmol, GC Std.). Hex-1-ene (15 mmol) was added slowly for 5 minutes at 0°C. The contents were further stirred for 2 h at rt. The reaction mixture was quenched with careful addition of water. The mixture was cooled to 10°C and 3N NaOH (3 mL) was added, followed by the slow addition of 30% H$_2$O$_2$ (2 mL) for 10 minutes. The contents were further stirred at 50°C for 2 h to ensure completion of the oxidation. The reaction mixture was cooled to room temperature and the organic layer was separated. The aqueous layer was saturated with K$_2$CO$_3$ and extracted with ether and combined organic extract was washed with brine and dried over anhydrous MgSO$_4$.

5 From Ethers

Ethers react with HI or HBr to give an alcohol and an alkyl halide. The iodide or bromide attacks the least substituted carbon of an unsymmetrical ether in an S$_N$2 reaction.

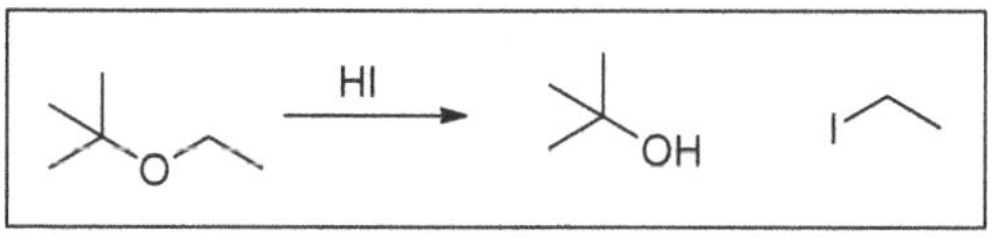

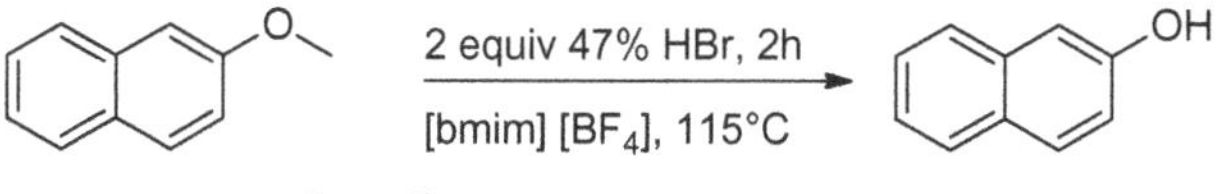

HI

2 equiv 47% HBr, 2h
[bmim] [BF$_4$], 115°C

J. Org. Chem. **2004**, 69, *3340–3444.*

Ether (1 mmol) and concentrated hydrobromic acid (47%, 2 mmol) in 1-*n*-butyl-3-methylimidazolium tetrafluoroborate, [bmim] [BF$_4$], 1.0 mL, were stirred at 115°C for an appropriate time. The reaction time was determined by TLC analysis. The reaction mixture was extracted with diethyl ether (4 × 10 mL). The combined ether extracts were concentrated under reduced pressure and the resulting product was purified by flash column chromatography to give the phenolic product. Note that [bmim] [BF$_4$] is an ionic liquid.

6 From Ketones or Aldehydes

Aldehydes and ketones are reduced to alcohols with NaBH$_4$ or LiAlH$_4$; with hydrogen gas and Pt, or Pd or Ni; with sodium in liquid ammonia and ethanol. Conjugated ketones and aldehydes react with NaBH$_4$ to primarily 1,2-reduction, or with LiAlH$_4$ to give a mixture of 1,2- and 1,4-reduction.

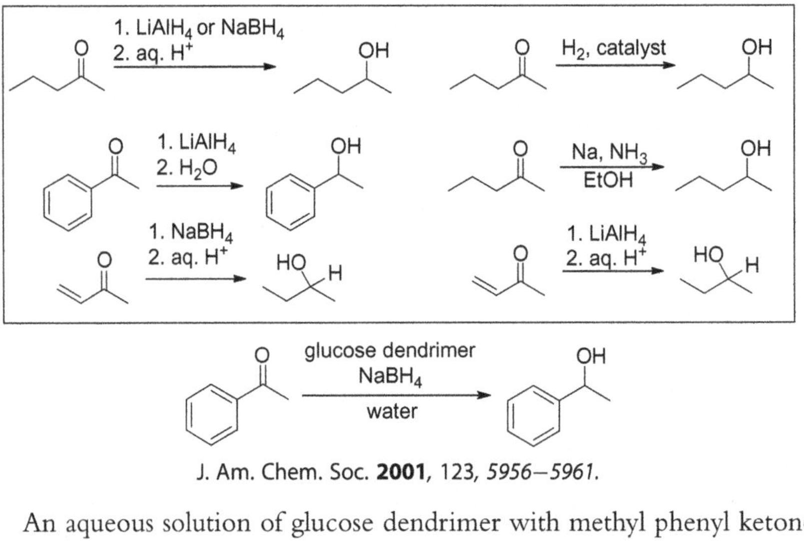

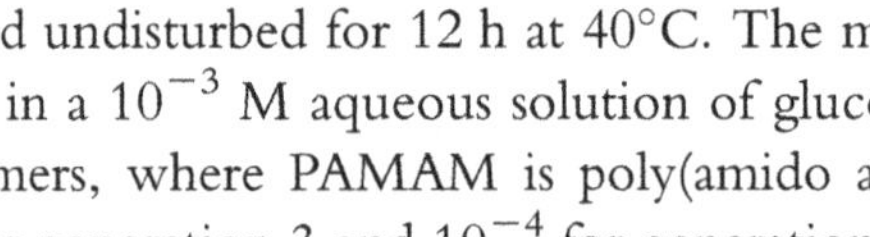

J. Am. Chem. Soc. **2001**, 123, *5956–5961.*

An aqueous solution of glucose dendrimer with methyl phenyl ketone in excess remained undisturbed for 12 h at 40°C. The maximum concentration of ketone in a 10^{-3} M aqueous solution of glucose-persubstituted PAMAM dendrimers, where PAMAM is poly(amido amine), was equal to 2 × 10^{-5} M for generation 3 and 10^{-4} for generation 4. Sodium borohydride was added to 10 mL of this saturated solution. The mixture was stirred for 2 h, followed by two successive extractions with carbon tetrachloride. After evaporation of the solvent, the methylphenylcarbinol was obtained and analyzed.

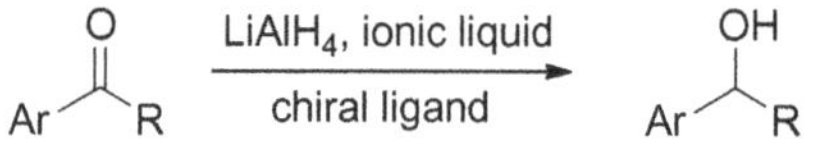

Tetrahedron: Asymmetry **2006**, 17, 1063–1065.

Reactions were carried out under a moisture-free nitrogen atmosphere. Ionic liquids were dried overnight in an oven at 70°C before each use. Chiral ligand (R)-1,1′bi-2-naphthol or (R)-6,6′dibromo-1-1′–bi-2-naphthol (2 mmol) was added to the ionic liquid (2 mL). The mixture was heated to 45°C and stirred until the solid dissolved. Lithium aluminum hydride (2 mmol) was slowly added to the mixture, which produced a very small amount of suspension. Furthermore, in this step, some bubbles were generated. The mixture was stirred (250 rpm) for 30 min at the specified reaction temperature and the aromatic ketone (2 mmol) was added dropwise. The reaction mixture was stirred (250 rpm) at this temperature for the desired time period. Finally 2 M HCl (5 mL) was added to quench the reaction mixture, which was then brought to room temperature. The organic compound was extracted by diethyl ether (5 mL), washed first with saturated sodium bicarbonate (5 mL), followed by brine (5 mL). Evaporation under reduced pressure yielded the concentrated organic mixture, which was further purified by flash chromatography (acetone/hexanes 1:7) to give purified products.

Alcohol-Alcohol
See Diols.

Alcohols-Functionalized
7 From Epoxides or Alkenes
Ethers react with nucleophiles at the less substituted carbon in non-aqueous solvents. Epoxides are reactive ethers that react with acids to form halo-alcohols. The reaction of an alkene with Cl_2/H_2O or Br_2/H_2O leads to a chlorohydrin or a bromohydrin.

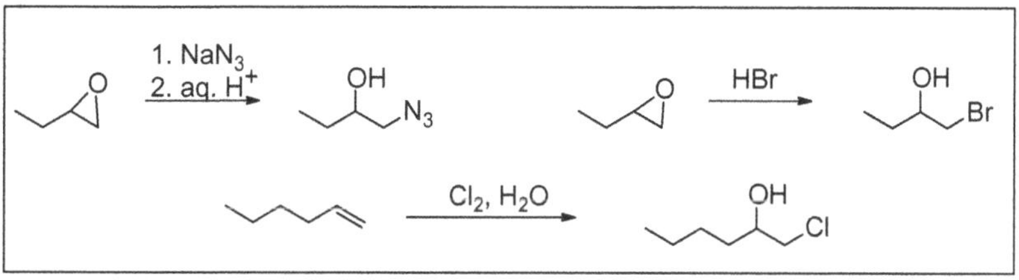

J. Am. Chem. Soc. **1980**, 102, 4193–4198.

Hydrogen bromide was slowly bubbled through a solution of styrene oxide (50 mmol) in chloroform (50 mL) at −15°C. After 2 h, the flask was capped and placed in a freezer overnight. The chloroform was then removed *in vacuo* and the residue was taken up in diethyl ether. After washing with water, 10% aqueous hydrochloric acid, and a saturated sodium bicarbonate solution and again with water, the extract was dried over magnesium sulfate. The ether was removed *in vacuo* and the residue was recrystallized from hexane to yield 2-bromo-2-phenylethanol (58%) as white crystals.

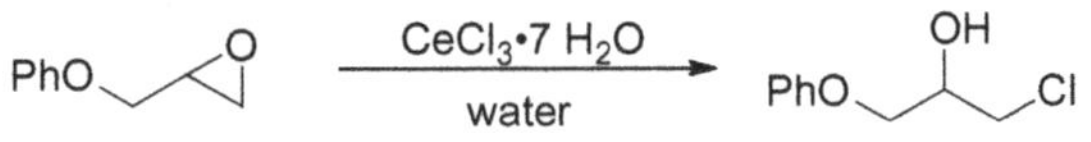

Tetrahedron Lett. **2001**, 42, 3955–3958.

To a solution of CeCl₃ · 7H₂O (0.5 mmol) and acetonitrile (10 mL) was added epoxide in acetonitrile (2 mL) and heated at reflux for 1 h. On completion, the solvent was removed under reduced pressure and extracted with ethyl acetate (2 × 10 mL) and washed with water and brine. After drying (Na₂SO₄) and solvent removal, the crude product was purified by column chromatography.

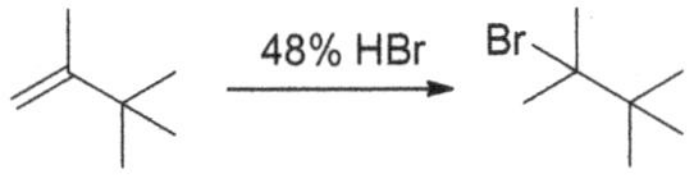

Organometallics **2010**, 29, 2735–2751.

2,3,3-Trimethylbutene (86 mmol) in dichloromethane (20 mL) was stirred vigorously with 48% HBr (25 mL) for 2 h. Phase separation, extraction with dichloromethane and drying over magnesium sulfate yielded 10.1 g (66%) of product, which was used without further purification.

Aldehydes or Ketones

8 From Acetals

Acetals react with aqueous acid to regenerate a ketone or aldehyde. Ketals are a sub-class of acetals, and they react with aqueous acid to regenerate a ketone.

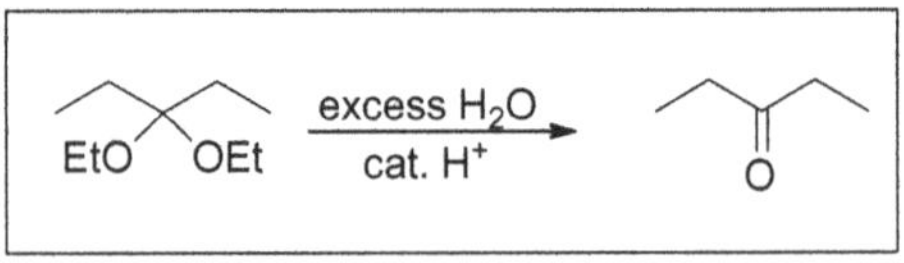

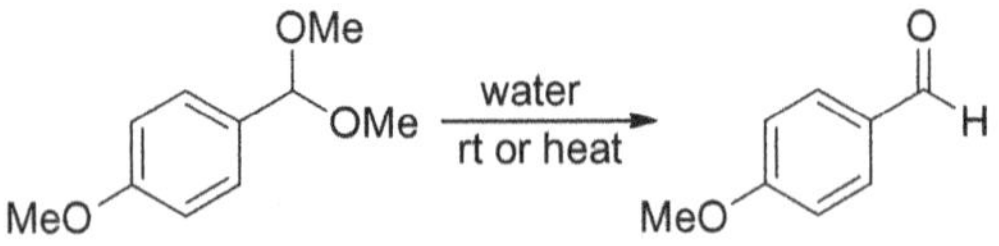

Green Chem. **2010**, 12, 1919–1921.

Deionized water (15 mL) was added to acetal (12.5 mmol). The reaction vessel was heated to 80°C for the determined period of time after which the water was simply removed by evaporation thereof. Alternatively, diethyl ether (3 × 5 mL) could be used with which to extract the organic material from the aqueous layer. The organic phase was dried with anhydrous magnesium sulfate and the volatile component was removed under vacuum.

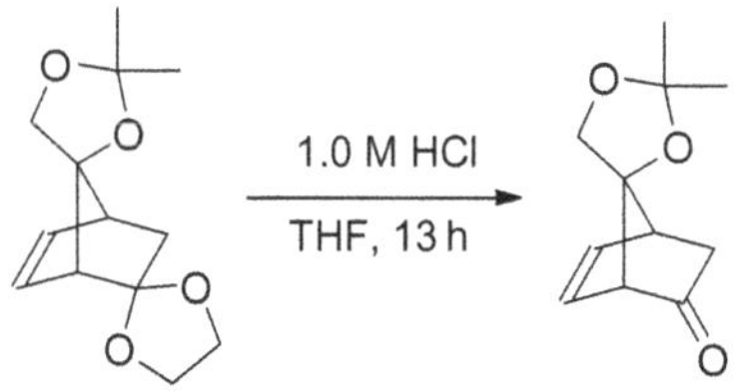

1.0 M HCl

THF, 13 h

J. Org. Chem. **1978**, 43, 4178–4182.

A solution of 3.49 g (14.7 mmol) of the ketal in 250 mL of tetrahydrofuran was cooled to 0°C and treated with 40 mL of 1 M hydrochloric acid. The reaction was warmed to room temperature and stirring was continued for 13 h. The reaction mixture was neutralized with 2 N solution of sodium hydroxide and the product was extracted with ethyl acetate. The combined organic layers were dried (MgSO₄), concentrated, and evaporated *in vacuo* leaving 2.72 g of a yellow oil, which was purified, on 260 g of silica gel. Elution with hexane-ether (3:2) gave 2.02 g (71%) of pure crystalline ketone.

9 From Alcohols

Primary and secondary alcohols are oxidized to aldehydes ketones using Jones reagent, Collins oxidation, PCC, or PDC. Primary alcohols are oxidized to aldehydes and secondary alcohols to ketones by Swern oxidation.

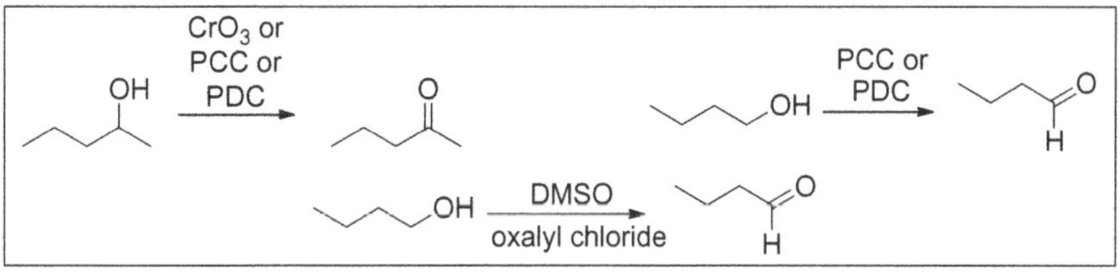

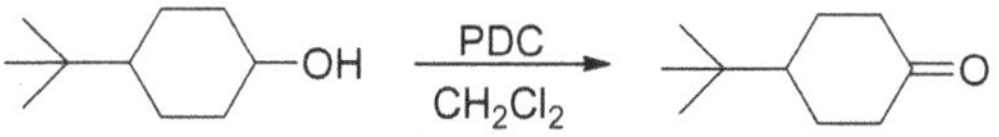

CrO₃ or
PCC or
PDC

PCC or
PDC

DMSO
oxalyl chloride

PDC
CH₂Cl₂

Tetrahedron Lett. **1985**, 26, 1699–1702.

4-*tert*-Butylcyclohexanol was stirred in CH_2Cl_2 (2 mL per gram PDC) with 1.5 equiv PDC and 0.4 eq. pyridinium trifluoroacetate at 25°C for 3 h. The mixture was then diluted with ether or ether-pentane and filtered. Last traces of Cr could be removed by filtering an ethereal solution through a small amount of anhydrous magnesium sulfate or silica gel.

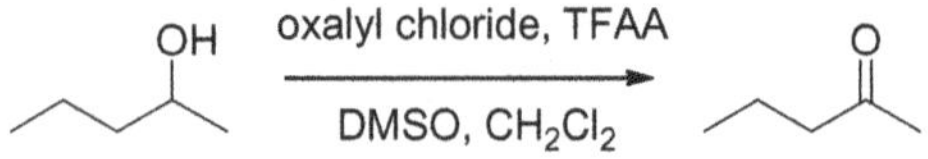

J. Org. Chem. **1976**, 41, *957–962.*

Methylene chloride (10 mL) and DMSO (20 mmol) were placed in a 50 mL three-neck flask equipped with a magnetic stirrer, thermometer, an addition funnel and a drying tube. The contents of the flask were cooled below −50°C with a dry ice-acetone bath and trifluoroacetic anhydride (TFAA, 15 mmol) in CH_2Cl_2 (∼5 mL) was added dropwise to the stirred cold solution in about 10 min. During the addition, a white precipitate of salt formed. After 10 min at −50°C, a solution of an alcohol (10 mL) in CH_2Cl_2 (5−10 mL) was added dropwise for about 10 minutes to the mixture maintained at -50°C. The reaction of the alcohol with the salt was exothermic. The mixture was stirred at −50°C for 30 minutes, followed by addition of triethylamine (TEA, 4 mL) dropwise in about 10 min. The contents of the flask were maintained at or below −50°C until addition of TEA was complete. The cooling bath was removed and the reaction mixture was allowed to warm to room temperature (about 40 min) and then subjected to GLC analysis.

10 From Alkenes

Alkenes are oxidatively cleaved to aldehydes, ketones or carboxylic acids with ozone, after reaction with dimethyl sulfide or hydrogen peroxide.

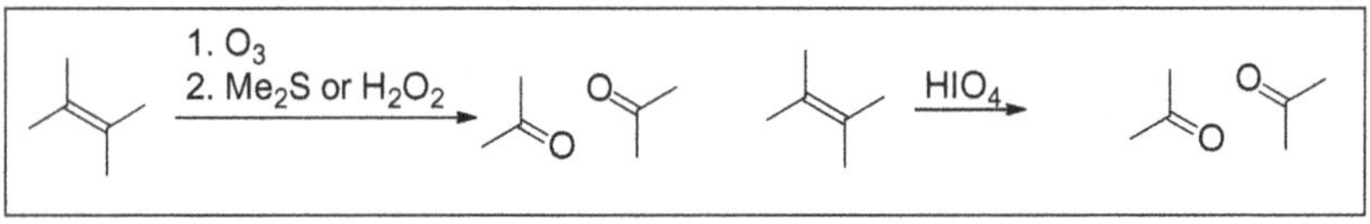

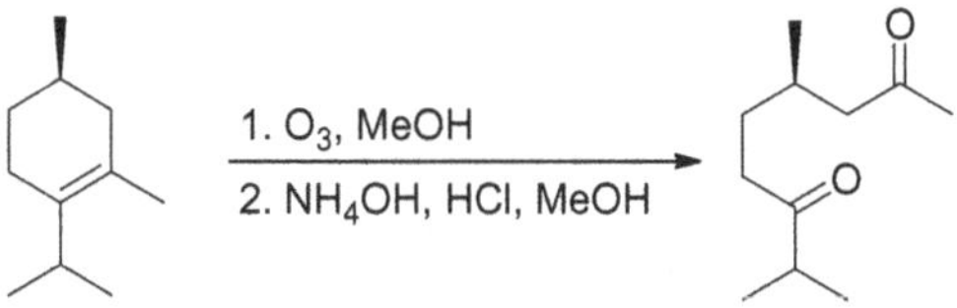

Chemistry of Natural Compounds **2006**, 42, *631−636.*

An O_3/O_2 mixture (calc 1 mol O_3 per 1 mol double bond) was bubbled through a solution of alkene (10 mmol) in absolute methanol (50 mL) at $0°C$. The reaction mixture was purged with Ar, stirred at $0°C$, treated over 0.5 h with $NH_4OH•HCl$ (35.0 mmol, per one double bond), boiled until the peroxide disappeared (10 h), evaporated, diluted with CH_2Cl_2 (100 mL) and washed with water. The organic layers were dried over Na_2SO_4 and the solvent evaporated.

11 From Alkynes

Hydration of alkynes with aqueous acid leads to an enol that tautomerizes to an aldehyde or a ketone; terminal alkynes react with boranes to give aldehydes, after treatment with $NaOH/H_2O_2$.

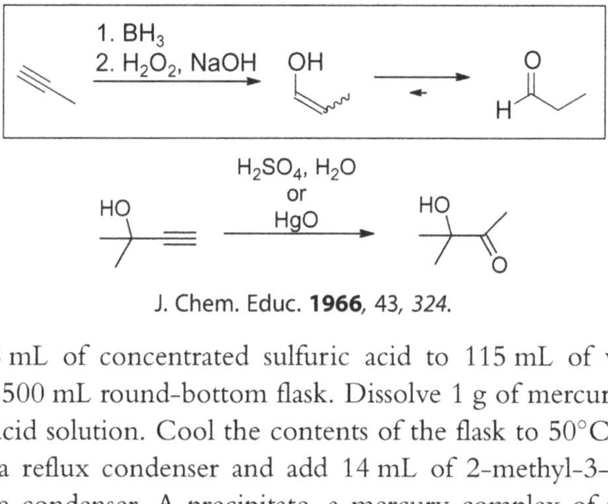

J. Chem. Educ. **1966**, 43, 324.

Add 18 mL of concentrated sulfuric acid to 115 mL of water contained in a 500 mL round-bottom flask. Dissolve 1 g of mercuric oxide in the warm acid solution. Cool the contents of the flask to $50°C$, equip the flask with a reflux condenser and add 14 mL of 2-methyl-3-butyn-2-ol through the condenser. A precipitate, a mercury complex of the alkyne, forms. Shake the flask to mix the contents. An exothermic reaction begins, the precipitate dissolves, and the solution turns a light brown. After the reaction is allowed to proceed by itself for two minutes, heat the reaction mixture to reflux. As soon as the reflux begins, stop the heating and cool the reaction mixture to $50°C$. Add a second 14 mL (a total of 26 g, 0.3 mol) of 2-methy1-3-butyn-2-ol. After a precipitate forms, mix the contents of the flask. After a few minutes, heat the contents of the flask to reflux, and reflux for 15 min. Transfer the reaction mixture to a 500 mL distilling flask, and add 100 mL of water to the reaction mixture and distill the reaction mixture. Collect a total of 150 mL of distillate. Add approximately 100 g of potassium carbonate sesquihydrate to the distillate to salt out the product. Separate the layers and extract the aqueous layer with 35 mL of benzene, dry the benzene solution with approximately 8 g of anhydrous potassium carbonate. Filter and distill. The desired product distills between 138 and 141°C. Average yield is 10−15 g of a clear, colorless liquid. Note: Benzene is a cancer suspect agent.

12 From Diols

Alkenes are oxidatively cleaved to aldehydes or ketones with periodic acid or with lead tetraacetate.

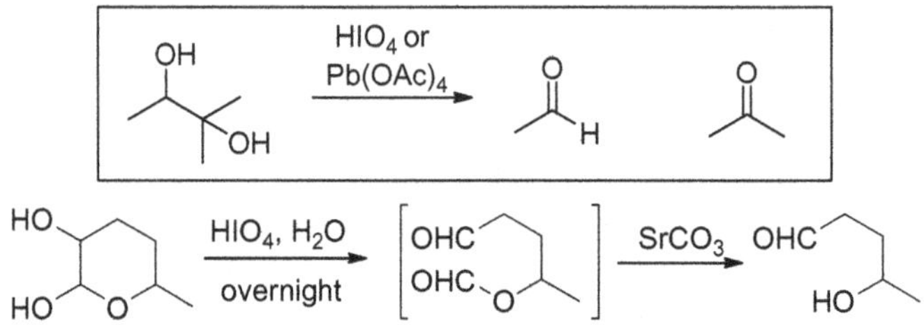

J. Org. Chem. **1958**, 23, 462−465.

A 2.0 g. (0.015 mol) sample of 6-methyltetrahydropyran-2,3-diol was allowed to stand overnight with 12.4 g of periodic acid in 120 mL of water. This was then made neutral to phenolphthalein with strontium hydroxide and filtered. About 0.8 g of strontium carbonate was added to the filtrate, which was concentrated to about 100 ml. volume at 70°C and 25 mmHg. This mixture was extracted with three 60-mL portions of ether, which were combined and dried. Removal of the ether left 0.5 g (32%) of 4-hydroxypentanal, which was distilled to make a final 26% yield of product.

13 From Nitriles

Nitriles are reduced to aldehydes with $SnCl_2$ and HCl.

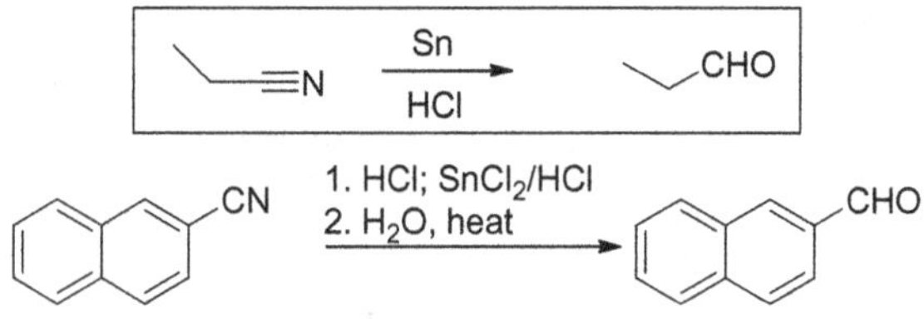

Organic Syntheses, Coll. Vol. 3 **1955**, 626.

In a 2 L. three-necked round-bottomed flask, provided with a mechanical stirrer, a reflux condenser carrying a drying tube, and an inlet tube reaching nearly to the bottom of the flask, was placed 76 g (0.4 mol) of anhydrous stannous chloride and 400 mL of anhydrous ether.

The mixture was then saturated with dry hydrogen chloride, while slowly stirred; this addition required 2.5–3 h, during which time the stannous chloride formed a viscous lower layer.

A dropping funnel replaced the inlet tube, and a solution of 30.6 g. (0.2 mole) of β-naphthonitrile (m.p. 60–62°C) in 200 mL of dry ether was added rapidly. Hydrogen chloride was again passed into the mixture until it was saturated, and the mixture was then stirred rapidly for 1 h and while standing overnight the yellow aldimine-stannic chloride separated completely. The ethereal solution was decanted, and the solid was rinsed with two 100-mL. portions of ether. The solid was transferred to a 5 L flask fitted for steam distillation and immersed in an oil bath, the temperature of which was maintained at 110–120°C. Dry steam was passed through the mixture until the aldehyde was completely removed; which required 8–10 h, and 8–10 L of distillate was collected.

The white solid was filtered and dried in the air to give 23–25 g (73–80%), which melted at 53–54°C. For further purification, it was distilled under reduced pressure, and the water-clear distillate (b.p. 156–158°C/ 15 mm.) was poured into a white-hot mortar and was pulverized when cool. The recovery was 93–95%, and the melting point is 57–58°C.

Alkanes

14 From Alkenes

Alkenes are reduced to alkanes with hydrogen gas and Pd, or Ni or Pt. Benzene is reduced to cyclohexane with an excess of hydrogen and a transition metal catalyst. Alkynes are reduced to alkanes with an excess of hydrogen gas and Pd, or Ni, or Pt.

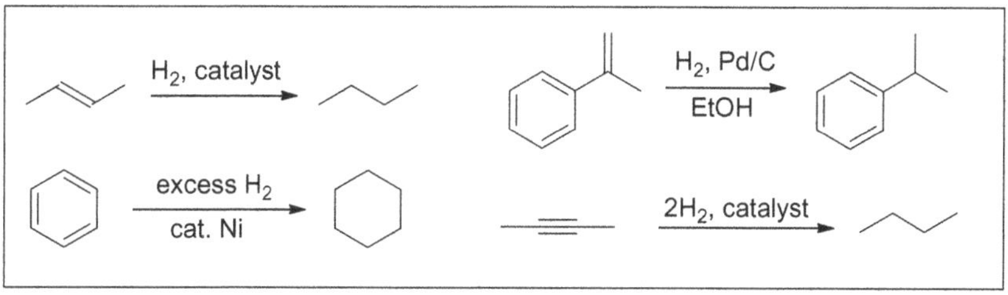

eg

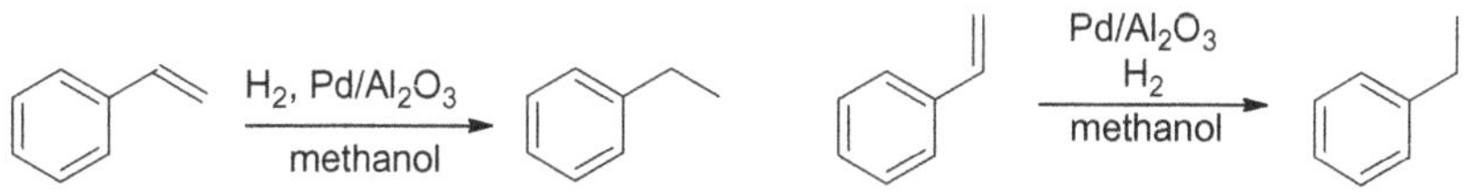

Synlett **2013**, 24, 2225–2228.

The hydrogenation was carried out in a 100 mL vial equipped with an MNB Generator without additional stirring (MNB — micro nanobubble). Alkene or alkyne (20 mmol) was dissolved in methanol (80 mL), and then warmed to 30°C. Using the MNB generator (MA3-FS), H_2-micro nano-bubbles were introduced into the reactor in the presence of palladium on alumina spheres (0.5% Pd, 2—4 mm, 0.1 mmol, 0.5 mol%) at a hydrogen flow rate of 5 mL/min. The samples of the reaction mixture were taken out periodically to monitor the reaction progress using the GC analysis. After the completion of the hydrogenation reaction, methanol was evaporated *in vacuo* to afford the desired alkane with excellent purity.

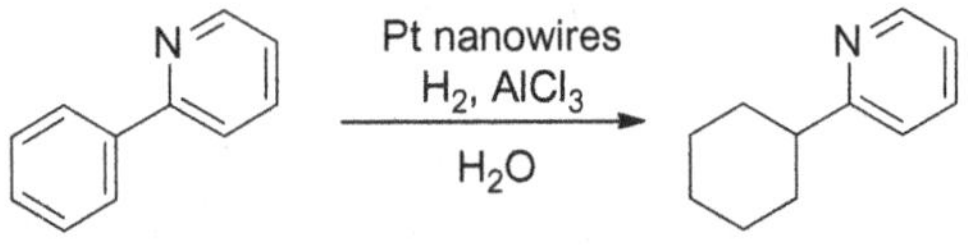

ChemCatChem **2013**, 5, 2852—2855.

Catalyst testing was performed in a sealed tube. Pt nanowires in ethanol were added and the ethanol was evacuated by pressure reducing. Aromatics, $AlCl_3$ and the solvents were added in the reaction tube and then sealed. The reaction tube was thrice evacuated and flushed with H_2 and took place at a certain temperature under hydrogen atmosphere.

15 From Alkyl Halides

Alkyl halides are reduced to the hydrocarbon with $LiAlH_4$; by reaction with Mg, followed by treatment with water; by reaction with Zn and HCl. Grignard reagents and organolithium reagents react with acids such as water, alcohols, and amines to form an alkane.

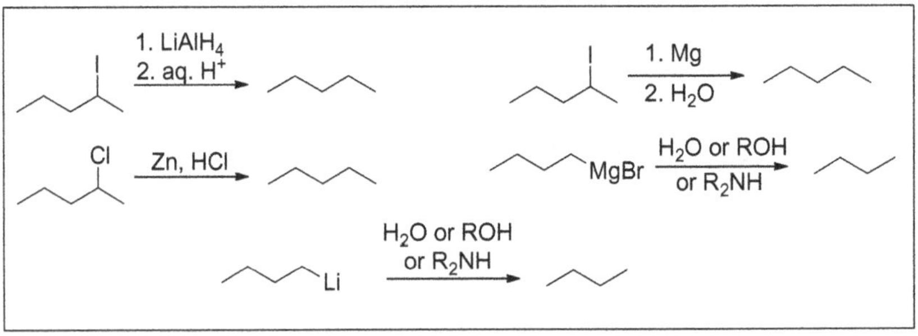

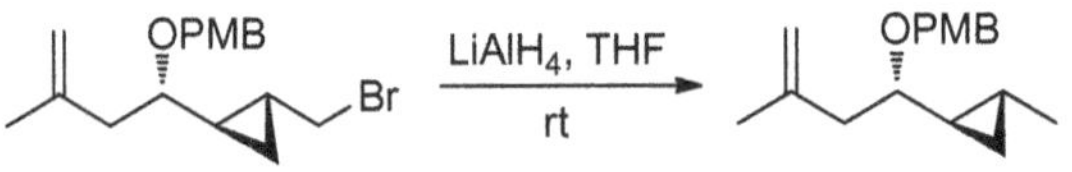

PMB = p-methoxybenzyl

Org. Lett. **2006**, 8, 661—664.

O-p-Methoxybenzyl-2-(bromomethyl)cyclopropyl)-3-methylbut-3-en-1-ol (1.68 mg, 4.95 mmol) in THF (4.4 mL) was added dropwise to the stirred solution of LiAlH$_4$ (7.43 mL, 7.43 mmol, 1M in THF) in THF (8 mL) at room temperature. The reaction mixture was stirred for 1 h, then was quenched at 0°C with THF/H$_2$O (1:1, 5 mL), followed by saturated aqueous sodium potassium tartrate (50 mL) and the contents were stirred at rt for 1 h. The aqueous material is extracted with ether (3×) and the combined organic extracts were washed with brine (1×), dried over MgSO$_4$, and then concentrated *in vacuo*. The residue was purified by flash column chromatography on silica gel (hexane:EtOAc = 10:1) to yield O-p-methoxybenzyl-3-methyl-1-(2-methylcyclopropyl)but-3-en-1-ol (1.23 g, 96%).

16 From Ketones or Aldehydes

Aldehydes or ketones are reduced to −CH$_3$ or −CH$_2$- with Zn/Hg and HCl (Clemmensen) or NH$_2$NH$_2$/KOH (Wolff-Kishner).

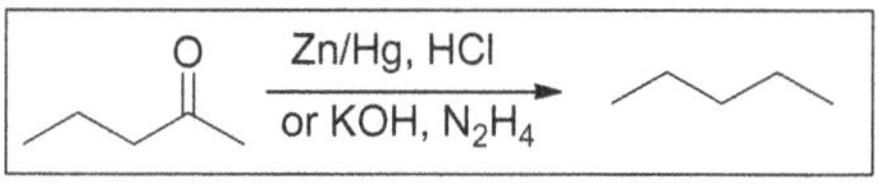

eg

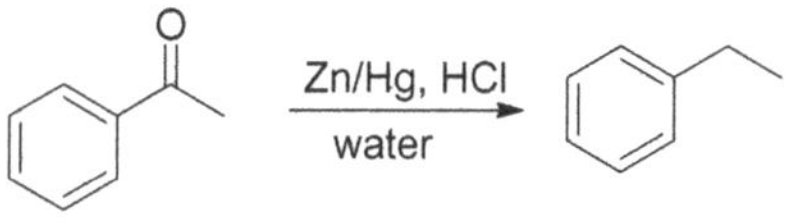

Chemistry and Industry **1965**, 65, 679.

Mossy zinc (20 g) was amalgamated (2.0 g HgCl$_2$) in the usual manner. Water (20 mL), ketone (10 g) and 37% hydrochloric acid (40 mL) were then added to the metal and the mixture was refluxed for 17 h. After this time, the mixture was cooled and then poured into water. The solution was then extracted twice with ether; this extract was washed with sodium bicarbonate and dried with sodium sulfate. The ether was removed by distillation and the hydrocarbon content was isolated and purified by distillation.

Alkenes

17 From Alkynes

Alkynes are reduced to alkenes with an one equivalent of hydrogen gas and Pd, or Ni or Pt; alkynes are reduced to (Z)-alkenes with an one equivalent of hydrogen gas and Pd/BaSO$_4$ and quinoline (Lindlar hydrogenation); Alkynes are reduced to (E)-alkenes with Na/NH$_3$ in ethanol (dissolving metal reduction).

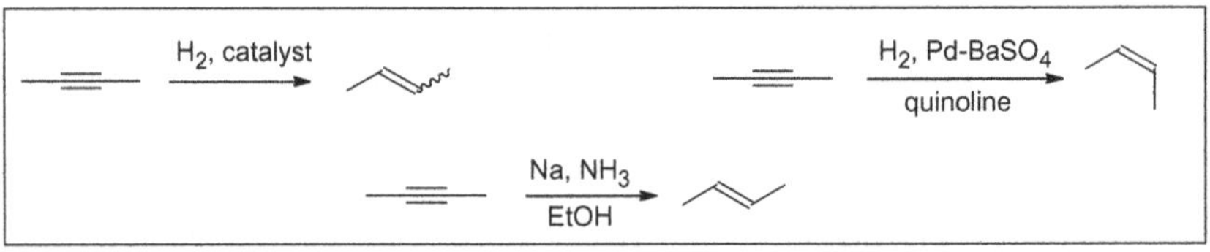

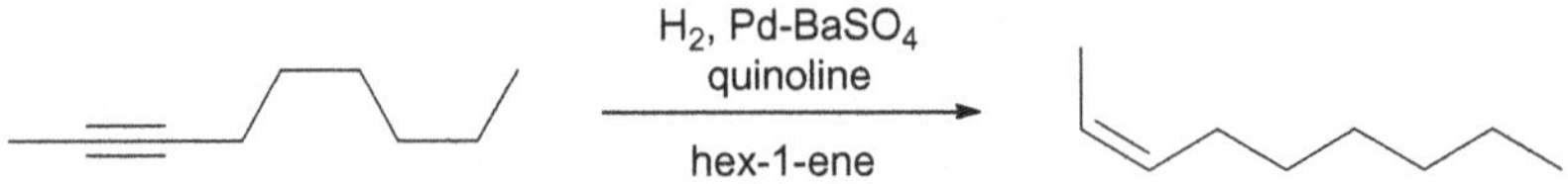

Org. Lett. **2009**, 5150–5153.

Quinoline (40.0 mmol) was added to a solution of non-2-yne (4.0 mmol) in hex-1-ene. The mixture was then degassed with Ar 3x and then 5% Pd/BaSO$_4$ (0.42 g) was added. The Ar was replaced with hydrogen and the mixture was stirred at room temperature under hydrogen atmosphere (balloon) while the reaction was closely followed by GC. After the starting material disappeared, the mixture was filtered through a pad of silica gel, concentrated and purified by flash column chromatography (hexanes) to give *cis*-non-2-ene as a colorless oil (0.33 g, 65%).

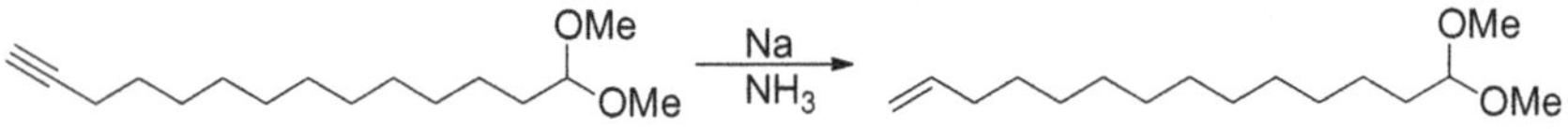

J. Org. Chem. **2003**, 68, 2820–2829.

Ammonia (60 mL) was condensed in a flask at −78°C and 14-methoxymethyl-1-tetradecyne (2.0 mmol) dissolved in THF (21 mL) was added. Sodium metal (9.8 mmol) was added. Once the sodium was dissolved, the mixture was maintained at -33°C for 8 h, the ammonia was evaporated and methanol (30 mL) was carefully added. The solvent was removed under vacuum and the crude was treated with saturated ammonium chloride solution (50 mL), extracted with ether (3 × 10 mL) and washed with brine (2 × 20 mL) and dried over magnesium sulfate and concentrated to give 401 mg (1.6 mmol, 80%) of 14-methoxymethyl-(E)-tetradec-2-ene as an oil.

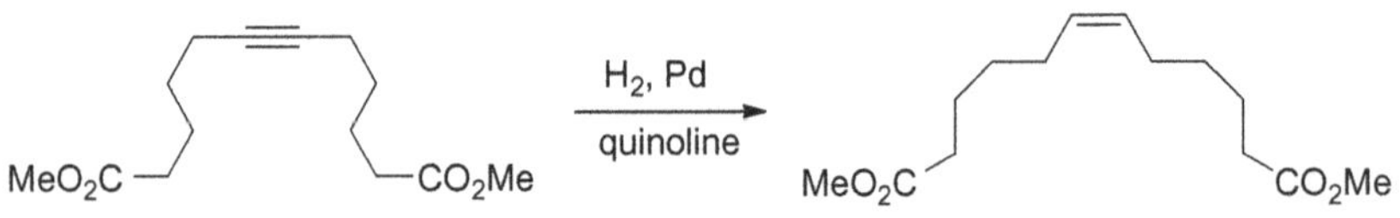

J. Am. Chem. Soc. **1956**, 78, 2518–2524.

The acetylenic diester (dimethyl dodec-6-ynedioate, 19.2 g), was hydrogenated using 0.4 g. of 5% palladium-on-barium sulfate and 0.4 g. of pure (synthetic) quinolone in 100 mL of methanol. The mildly exothermic reaction ceased abruptly after 20 minutes, exactly one equivalent of hydrogen having been taken up. The catalyst was removed by filtration, and after distilling the solvent and quinolone, the ester was distilled, to give 18.9 g of dimethyl (Z)-dodec-6-enedioate (97%).

18 From Halides, Sulfonate Esters, or Ammonium Salts

Alkyl halides and alkyl sulfonate esters undergo elimination reactions with a strong base to give an alkene (E2). Trialkylammonium hydroxides give the less substituted alkene upon heating (Hoffman Elimination).

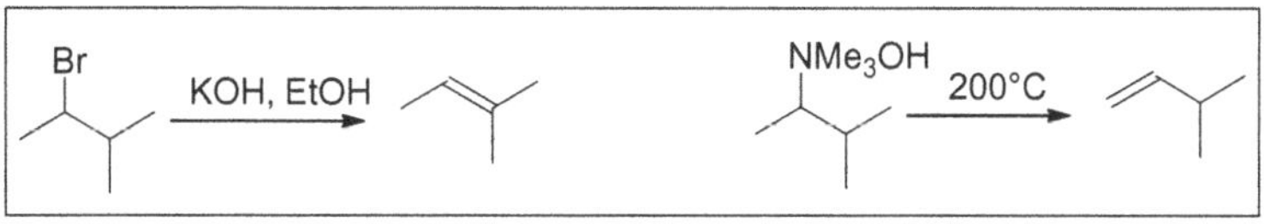

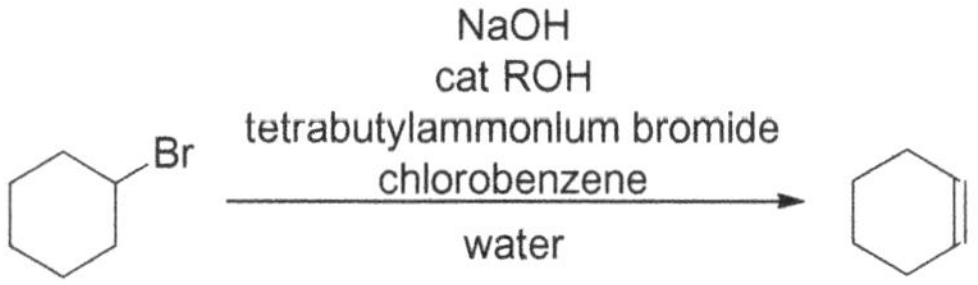

Tetrahedron **2000**, 56, 3553–3558.

A homogenous mixture of bromocyclohexane (20 mmol), chlorobenzene (2.00 g), tetrabutylammonium bromide (1 mmol) and an alcohol cocatalyst (1 mmol) was rapidly added to 50% aqueous solution of NaOH (~190 mmol) thermostated at 40°C and stirring was turned on. The reaction was stirred for 45 min at ~40°C and then removed from the heating bath. Dichloromethane (25 mL) was added to the reaction mixture and the separated organic layer was washed with water (3 × 5 mL), dried over magnesium sulfate and analyzed by GC using chlorobenzene present in the mixture as the internal standard.

Alkyl Halides

See Halides, Alkyl.

Alkynes

19 From Alkenyl Halides

The reaction of a vinyl halide with a strong base gives an alkyne (E2).

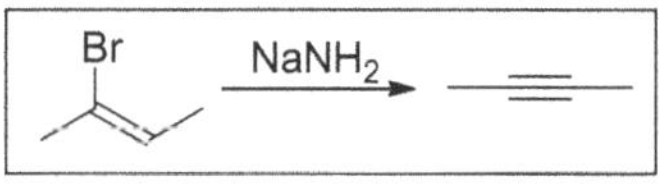

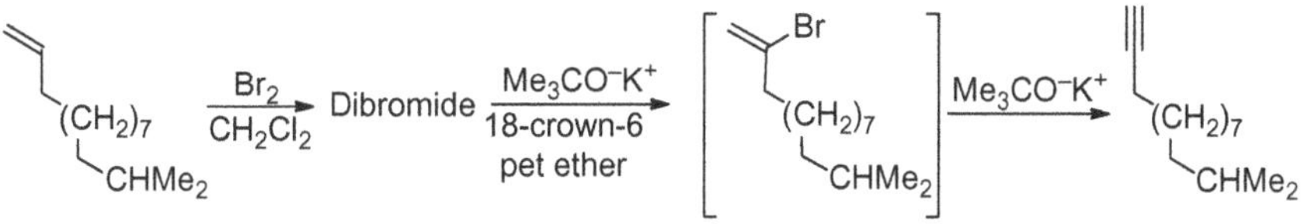

J. Chem. Soc. Perkin Trans. 1 **1999**, 2467–2477.

To a solution of 12-methyltridec-1-ene (2.71 g, 13.8 mmol) in dry CH_2Cl_2 (30 mL), bromine (0.78 mL, 15.2 mmol) was added and the mixture was stirred for 10 min at 0°C. After quenching with saturated aq. $Na_2S_2O_3$, it was extracted with n-hexane. The extract was washed with water and brine, dried ($MgSO_4$), and concentrated under reduced pressure to give 1,2-dibromo-12-methyltridecane (5.00 g, quantitative) as a colorless oil. This was employed in the next step without further purification. To a solution of the dibromide (5.00 g, 14.0 mmol) in petroleum ether (70 mL), KOtBu (4.65 g, 41.4 mmol) and 18-crown-6 (11 mg, 0.041 mmol) were added and the mixture was stirred for 2 h at reflux. This mixture was poured into water and extracted with n-hexane. The extract was washed with dilute aq HCl, water, and brine, dried ($MgSO_4$), and concentrated under reduced pressure. The residue was chromatographed on SiO_2 and distilled to give 12-methyltridec-1-yne (1.93 g, 72%), as a colorless oil. It is likely that initial base-induced elimination of the dibromide gave a vinyl bromide, which reacted with additional base to give the alkyne.

Amides

20 From Acid Derivatives

Carboxylic acids can be converted to the amide by reaction with an amine and DCC or reaction with an amine followed by heating. Acid chlorides, acid anhydrides, and esters react with amines or ammonia to give an amide.

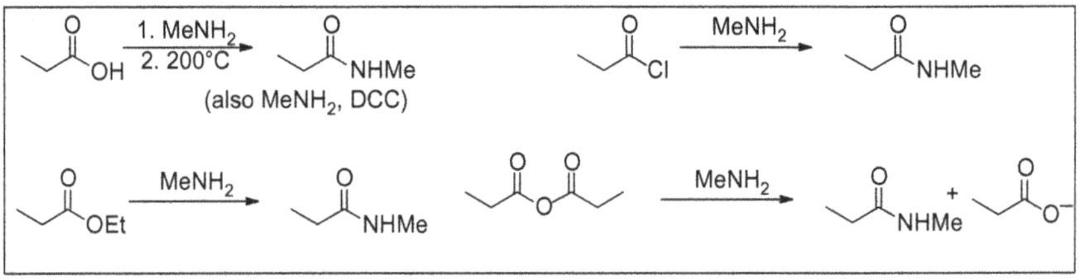

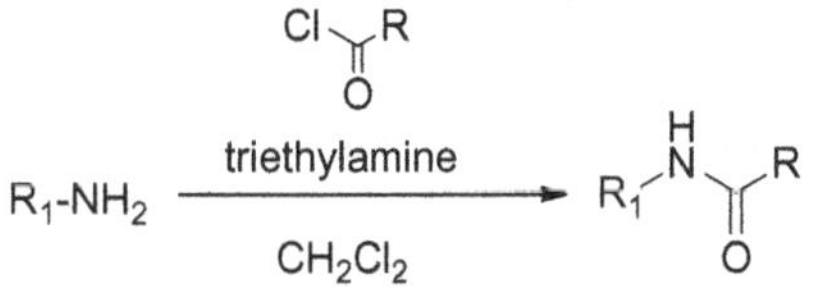

J. Am. Chem. Soc. **2012** 134, 8298–8301.

To a round-bottom flask equipped with a magnetic stir bar were added under argon the amine (1.10 equiv), triethylamine (1.25 equiv) and dichloromethane. The acyl chloride (1.00 equiv) was added in one portion and the reaction mixture was stirred at room temperature until completion (typically 30 minutes). The mixture was diluted with dichloromethane and washed successively with 1M NH_4Cl and brine. The organic layer was dried over $MgSO_4$, filtered and concentrated under reduced pressure. The residue was purified by flash column chromatography on silica gel.

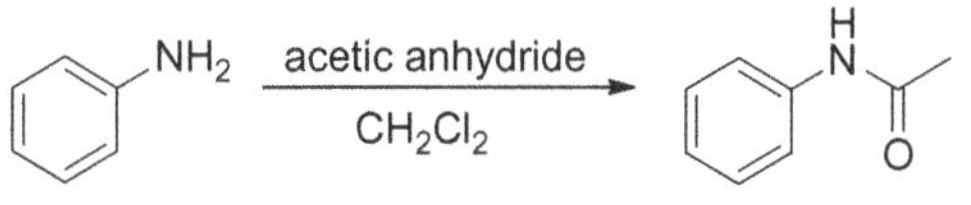

J. Am. Chem. Soc. **2008**, 130 *16474–16475*.

Aniline (109.7 mmol) was added to a round-bottom flask via syringe and fitted with a rubber septum. The flask was purged with argon and dry methylene chloride (300 mL) was added. Acetic anhydride (132.2 mmol) was added and the reaction was stirred at room temperature and monitored by TLC. Upon completion (generally a couple of hours but as short as 20 minutes) the reaction mixture was washed with a saturated solution of sodium carbonate, the organic layers dried with $MgSO_4$ and the solvent removed under reduced pressure. The product was obtained in quantitative yield.

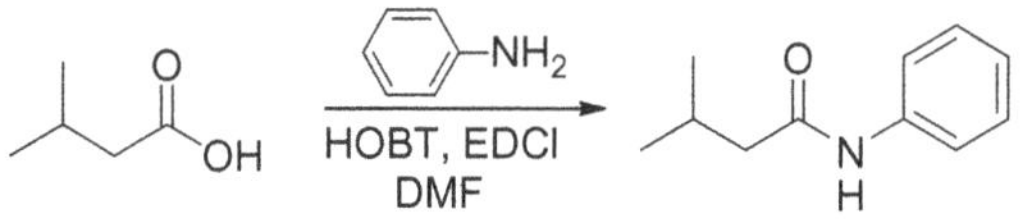

J. Curr. Pharm. Res. **2012**, 10, *22–24*.

In a round-bottom flask under a nitrogen atmosphere, isovaleric acid (0.49 mmol) was dissolved in DMF (5 mL). To this solution, EDCI [1-ethyl-3-(3-dimethylaminopropyl)carbodiimide, 0.54 mmol] and HOBT (hydroxybenzotriazole, 0.59 mmol) were added. The reaction mixture was stirred at rt for 15 min and then the solution of aniline (0.59 mmol) in DMF (5 mL) was added to this reaction mixture. The reaction was monitored using TLC. To this reaction mixture, ethyl acetate (20 mL) was added. It was washed with 10% aqueous solution of sodium bicarbonate. Two layers were separated properly. The ethyl acetate layer was washed with 10% HCl solution and two layers were separated. It was then washed with 10 mL brine solution. The ethyl acetate layer was dried over anhydrous sodium sulfate and evaporated in vacuum to give coupled compound. It yielded crude product, which was purified by recrystallization from methanol.

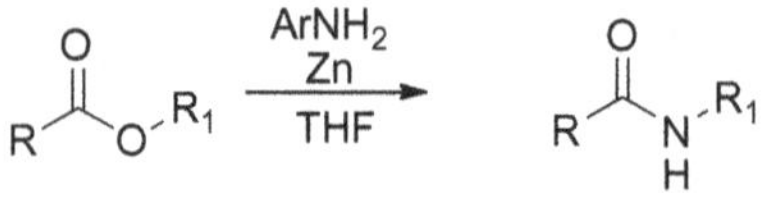

Can. J. Chem. **2005**, 83, 1137–1140.

To a mixture of the aromatic carboxylic ester (5 mmol), the appropriate aromatic amine (5 mmol) and Zn dust (2.5 mmol) in a round-bottom flask (50 mL), THF (10 mL) was added. The round-bottom flask was fitted with a reflux condenser and the reaction mixture was stirred at 70°C in an oil bath for an appropriate time. After completion as indicated by TLC, the reaction mixture was cooled and filtered. The THF was removed under reduced pressure and the product was obtained by filtration, after addition of DMF (5 mL) to the reaction mixture. Ice-cold water (100 mL) was added to the filtrate and the product extracted with ethyl acetate (3 × 15 mL). The combined extracts were washed with water and dried over anhydrous sodium sulfate. The product was obtained after the removal of the solvent under reduced pressure followed by crystallization from ethyl acetate-pet ether or by passing through a column of neutral alumina and elution with pet ether.

21 From Nitriles

Nitriles react with aqueous acid or base to give a carboxylic acid or an amide.

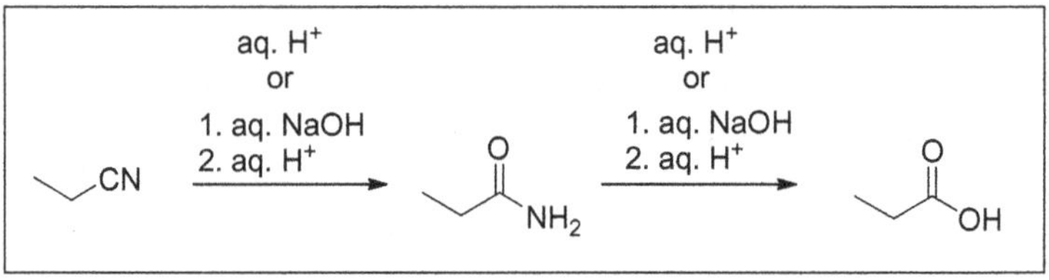

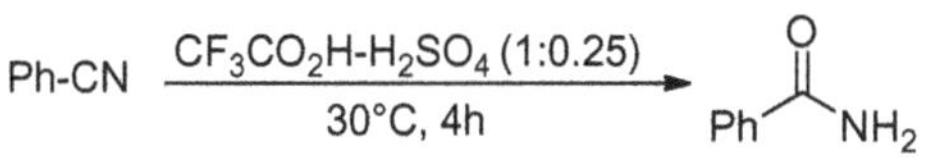

J. Org. Chem. **2005**, 70, 1926–1929.

A solution of 5.0 mmol of benzonitrile in a ≈7.0 mL mixture of trifluoroacetic acid-H_2SO_4 (4:1, v/v) was stirred at 30°C. The progress of the reaction in each case was monitored by TLC analysis. After completion of the reaction, the reaction mixture was poured into ice-cold water and the product filtered to give benzamide in 87% yield.

Amines

22 From Azides, Imides, and Halides

Alkyl azides, prepared from alkyl halides, are reduced to the corresponding amine; phthalimide derivatives, prepared from alkyl halides, are hydrolyzed to the amine or treated with hydrazine to give the corresponding amine. Halobenzene derivatives with electron releasing substituents react with nucleophiles via a benzyne intermediate to give substitution products.

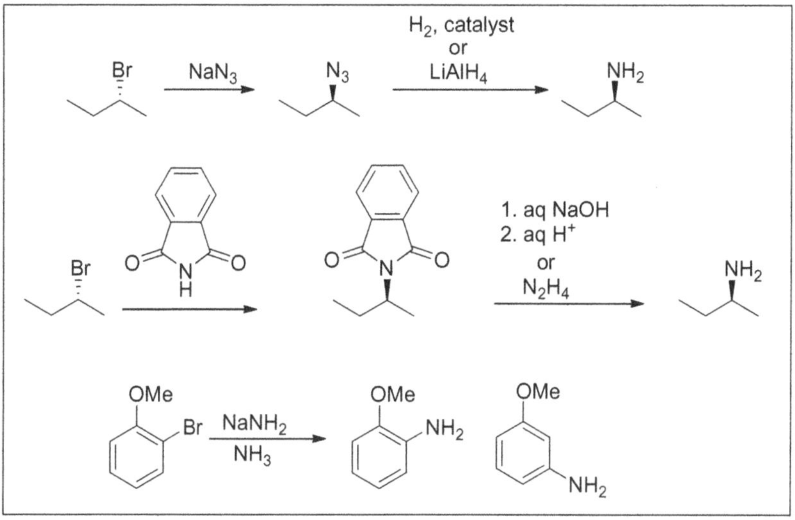

Eur. J. Org. Chem. 2004, 1732–1739.

A solution of 5-azido-4-benzyl-2,2-dimethyl-1,3-dioxane (1.0 equiv.) in THF (3 mL/mmol) was added dropwise to a stirred suspension of LiAlH$_4$ (2.0 equiv.) in THF (3 mL/mmol) at room temp. under argon. The reaction mixture was stirred for 20 min and then a 10% aq NaOH solution (3.6 equiv) was added dropwise, and diluted with CH$_2$Cl$_2$. This biphasic system was stirred for 5 min and then separated. The aqueous phase was further extracted with CH$_2$Cl$_2$ and the combined organic phases were washed with brine and dried (MgSO$_4$). Concentration under reduced pressure afforded 4-benzyl-2,2-dimethyl-1,3-dioxan-5-amine as a colorless oil. Yield: 160 mg (99%).

23 From Nitriles

Nitriles are reduced to the amine with LiAlH$_4$ or via catalytic hydrogenation.

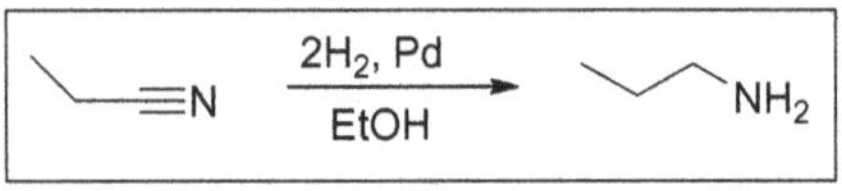

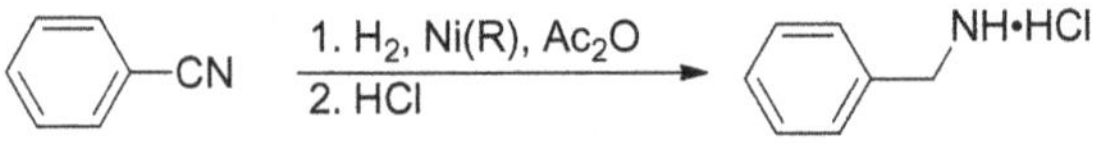

J. Org. Chem. **1960**, 25, 1658–1660.

A mixture of treated Raney Nickel (2–3 g), benzonitrile (0.1 mol) and acetic anhydride (120 mL) was shaken at 50°C under an initial hydrogen pressure of 50 psi. When the reaction was complete (1 h), the mixture was filtered hot and the filtrate was treated with water (40 mL). Then conc HCl (180 mL) was added and the mixture was heated under reflux for 16 h. The resulting solution was cooled to 25°C, made strongly basic with 5 N NaOH solution, extracted with ether (2 × 100 mL). The ether solution was dried over anhydrous magnesium sulfate, filtered and treated with gaseous hydrogen chloride until no further precipitate formed. After recovery and drying, there was obtained 13.0 g of benzylamine hydrochloride.

24 From Nitro Compounds

Nitro units attached to a benzene ring are reduced to give aniline derivatives.

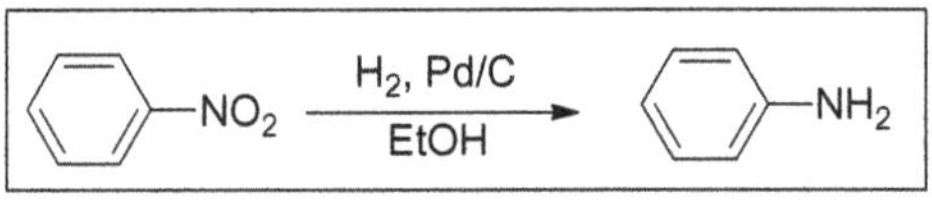

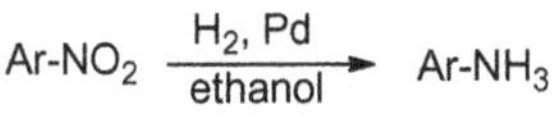

ACS Nano **2014**, 8, 5352–5364.

Nitroarene (0.5 mmol), Pd-containing samples (0.5 mol%) and ethanol (2.0 mL) were placed in a Schlenk flask (20 mL). The flask was purged with H$_2$ three times to remove air and the reaction mixture was stirred with a balloon of H$_2$ at room temperature for a given time. After the reaction, the resultant mixture was transferred into a tube and the solid was separated by centrifugation.

Anhydrides
25 From Carboxylic Acids and Acid Derivatives
Carboxylic acids react with other carboxylic acids to give an acid anhydride; acid chlorides and esters react with carboxylic acids to give an acid anhydride.

1. ZnO, toluene
2. RC(O)Cl

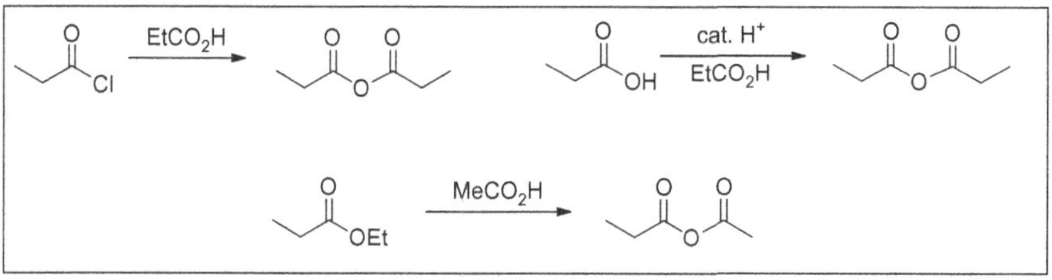

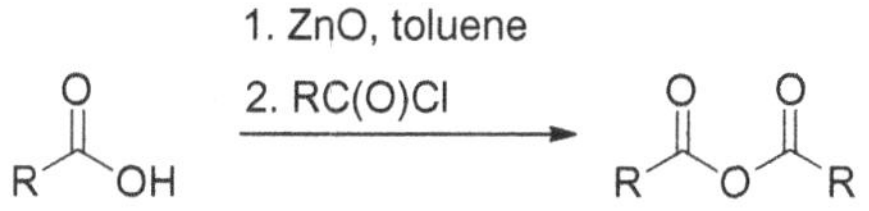

Indian J. Chem. **2005**, 44B, 420–421.

Zinc oxide (5 mmol) and the carboxylic acid (10 mmol) were magnetically stirred in toluene (25 mL), and then heated at reflux for about 2–4 h in a flask equipped with a Soxhlet extractor filled with anhydrous sodium sulfate to remove the water of dehydration by azeotropic extraction. After the completion of the reaction, carboxylic acid chloride (10 mmol) was added while stirring. The mixture was stirred at 40°C till the reaction was complete. The reaction mixture was filtered, washed with ice-cold 5% sodium bicarbonate (10 mL) and extracted with ether (2 × 10 mL). The combined extract was dried over anhydrous sodium sulfate and the solvent removed under vacuum to get the anhydride.

Aryl Halides

Alkyl. See Halides, Aryl.

Carboxylic Acids
26 From Acid Derivatives (acid chlorides, anhydrides, esters, amides)
Acid chlorides, acid anhydrides, esters, and amides react with aqueous acid or base to give a carboxylic acid.

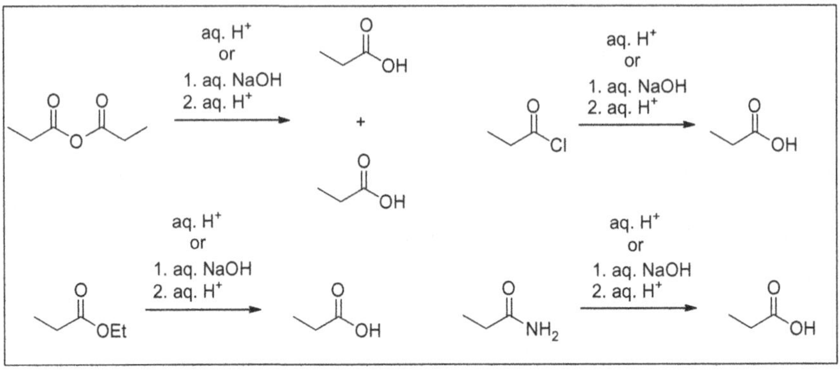

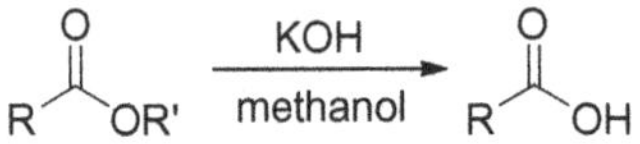

Monatshefte Für Chemie, **2004**, 135, 83–87.

In a typical experiment, 1 g of ester and 3 mL of methanol were placed in a 25 mL round-bottom flask mounted over a magnetic stirrer and maintained at ∼35°C. Potassium hydroxide was added and the contents stirred. The reaction was quenched invariably after 60 min by the addition of 10 mL of water. Unreacted ester, if any, was removed by ether extraction (2 × 5 mL) and the carboxylic acid was recovered by acidification to pH=2 with 6 N HCl and extraction into ether (3 × 5 mL). The combined ether extracts were dried with Na_2SO_4 and concentrated on a Büchi rotary evaporator.

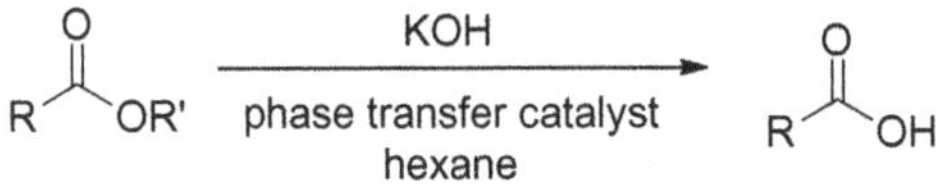

Org Prep. Proceed. Int. **2003**, 35, 361–368.

Saponification of esters was performed under microwave irradiation or conventional heating. In a Pyrex cylindrical open reactor adapted to the synthewave reactor (a microwave reactor), ester (5 mmol) was mixed with base (10 mmol), Aliquat 336 (0.5 mmol), and 5 mL of hexane when appropriate) The mixture was then exposed to microwave irradiation under mechanical stirring. At the end of irradiation, the reaction mixture was cooled to room temperature and extracted with methylene chloride. The aqueous layer was acidified with a solution of 2 N HCl (3 mL) and extracted with methylene chloride. The organic layer was dried over magnesium sulfate and concentrated under vacuum to give a white solid requiring no further purification.

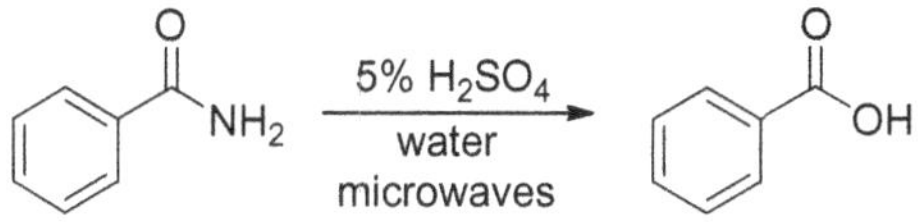

ChemSusChem **2008**, 1, *123−132*.

A mixture of benzamide (8.3 mmol) and 5% (v/v) sulfuric acid (10 mL) was filled in a large Biotage microwave process vial (20 mL). The vial was sealed and irradiated for 7 min (fixed hold time) at 180°C with magnetic stirring. After cooling to 48°C with compressed air, the mixture was allowed to stand in the refrigerator (4°C) for 3−4 h to enable complete crystallization of benzoic acid. The obtained solid was isolated by filtration and washed with ice-cold water and subsequently dried overnight at 50°C to give benzoic acid in 90% yield.

27 From Alcohols

Primary alcohols can be oxidized to a carboxylic acid using Jones oxidation.

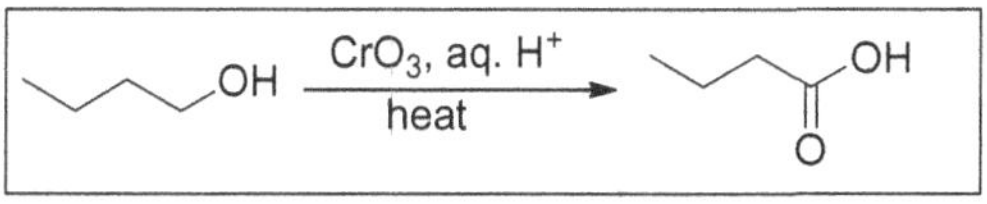

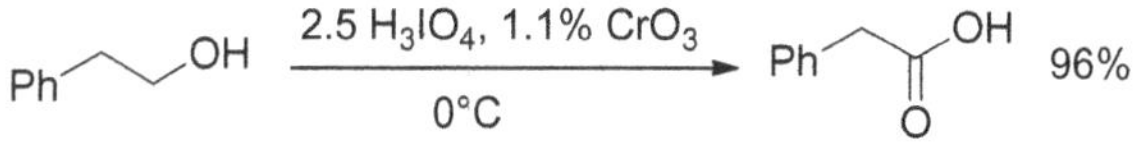

Tetrahedron Lett. **1998**, 39, *5323−5326*.

A stock solution of H_3IO_4/CrO_3 was prepared by dissolving H_3IO_3 (50 mmol) and CrO_3 (1.2 mol%) in wet acetonitrile (0.75 v% water) to a volume of 114 mL (complete dissolution typically required 1−2 hours). The H_3IO_4/CrO_3 solution (11.4 mL) was then added to a solution of 2-phenylethanol (2.0 mmol) in wet acetonitrile (10 mL, 0.75 v% water) in 30−60 minutes while maintaining the reaction temperature at 0−5°C. The mixture was aged at 0°C for 0.5 h and the completion of the reaction was confirmed by HPLC assay. The reaction was quenched by addition of an aqueous solution of Na_2HPO_4 (0.6 g in 10 mL H_2O). Toluene (15 mL) was added and the organic layer was separated and washed with 1/1 brine/water mixture (2 × 10 mL) then a mixture of aqueous $NaHSO_3$ (0.22 g in 5 mL water) and finally brine (5 mL). The organic layer was then concentrated to give phenylacetic acid.

28 From 1,3-Dicarboxylic Acids

1,3-Diacids undergo decarboxylation when heated to give a mono-carboxylic acid.

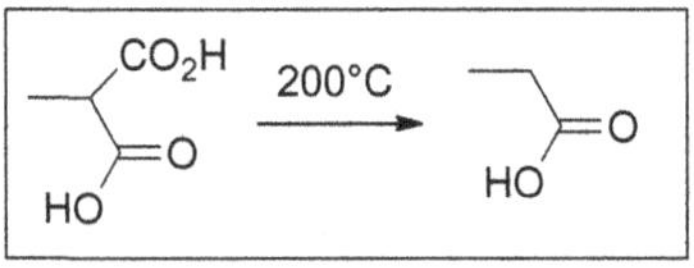

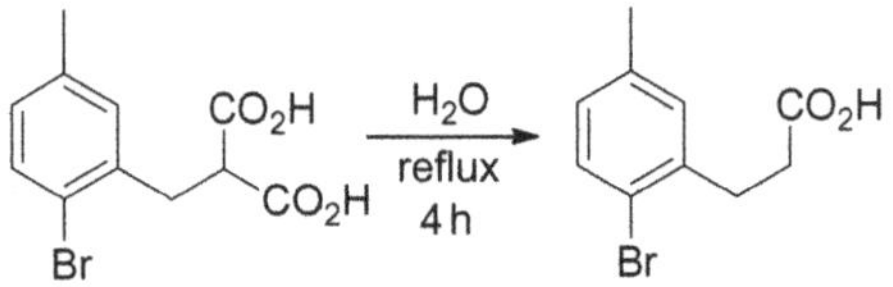

The most satisfactory results were obtained by boiling a solution of the malonic acid mixture (100 g) in water (400 mL) for four hours. The methylbromophenylpropionic acid mixture soon separated from the hot solution as an oil, and after cooling, the product was recovered by ether extraction. No alkali-insoluble material was present, and the acid mixture boiled at 168–172°C at 2 mm.; yield, 78 g (92%). This material, as well as that obtained from either of the pure malonic acids, was a very viscous oil at room temperature.

In an early experiment the decarboxylation was accomplished by heating the malonic acid mixture at 190°C, but the desired acid was obtained in only 70% yield and it was accompanied by a neutral substance, b.p. 129°C at 1.5 mm. This neutral substance proved to be a mixture of two esters, the methylbromophenylpropionic acid ethyl esters, C_6H_3Br-$(CH_3)CH_2CH_2COOC_2H_5$. Upon hydrolysis and ring closure, these esters gave a mixture of the two hydrindones (in the usual ratio).

29 From Nitriles

Alkyl halides react with KCN or NaCN to give a nitrile, which can be hydrolyzed to an amide or a carboxylic acid.

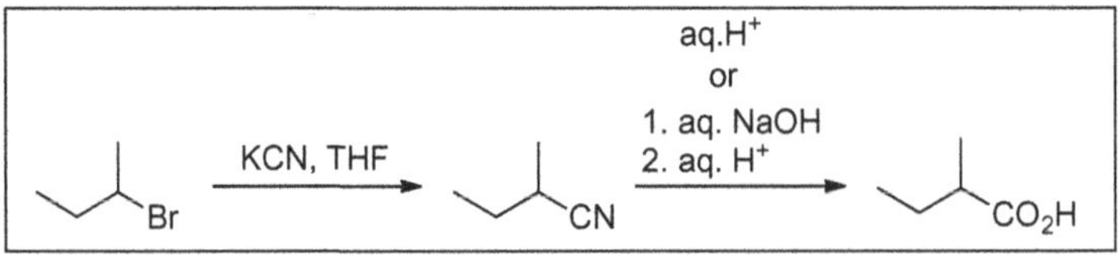

Hydrolysis by base

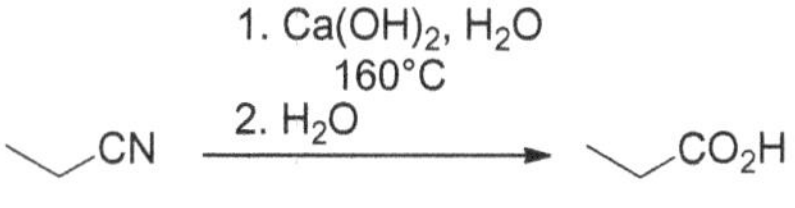

$$\text{1. Ca(OH)}_2\text{, H}_2\text{O}$$
$$160°\text{C}$$
$$\text{2. H}_2\text{O}$$

CN $\xrightarrow{\hspace{2cm}}$ CO$_2$H

Patent: US 2004/0002618 A1.

Propionitrile (0.4 mol), calcium hydroxide (0.31 mol) and deionized water (4.54 mol) were charged to a 350 mL, 316 stainless steel autoclave. The autoclave was closed, agitation was initiated and the unit was pressurized to 106 psi with helium. A helium flow of 30 ml/min through the head space was established. The contents of the autoclave were heated with an electric mantel to 161°C, over a period of 0.5 hour. The autoclave temperature was then controlled at 160−161°C while the autoclave pressure was maintained at 127−142 psi. After 1 hour at 161°C ammonia was detected in the gas leaving the autoclave with moist pH paper. The heat was turned off after 6.8 hours at which time ammonia was still detectable in the gas leaving the autoclave. The autoclave was cooled overnight and opened the next morning. The autoclave contained a white slurry with a very strong ammonia odor. A sample of the reaction mixture was taken, diluted 1:17.3 with deionized water and analyzed by HPLC. The mixture was found to contain 24.2% calcium propionate.

Hydrolysis by acid

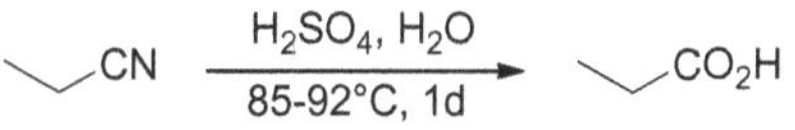

CN $\xrightarrow[\text{85-92°C, 1d}]{\text{H}_2\text{SO}_4\text{, H}_2\text{O}}$ CO$_2$H

Patent: 2007/0161820 A1.

A 100 mL three-neck round-bottom flask was set up with a water-cooled reflux condenser, an addition funnel and a thermometer. With magnetic stirring, propionitrile and DI water were added into the flask and concentrated sulfuric acid was added slowly into this mixture using the addition funnel. The reaction was then held at 85−92°C for 24 hours, and the reaction was allowed to cool to room temperature overnight, followed by separation of the reaction phases, which were then weighed and assayed by GC.

Dienes

30 From Aromatics

Benzene is reduced to cyclohexa-1,4-dienes with sodium in liquid ammonia and ethanol.

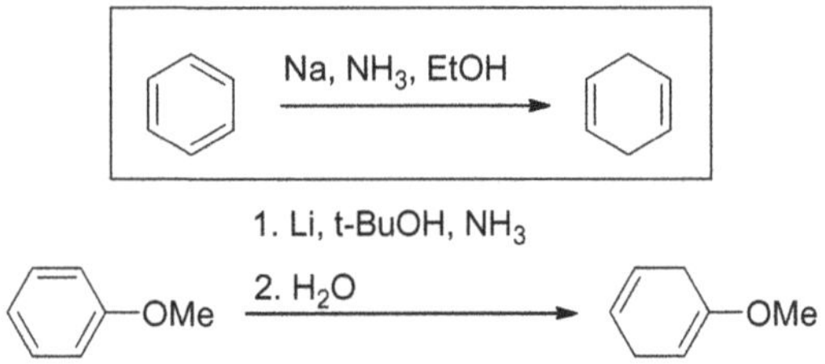

Org. Biomol. Chem. **2010**, 8, *539–545.*

Under nitrogen, anisole (0.19 mol) was dissolved in THF (50 mL) and *t*-BuOH (80 mL) and cooled to −78°C. Liquid ammonia (500 mL approx.) was then introduced. Lithium metal (0.53 mol) was added in portions to the mixture. Stirring was continued for 1 h at -33°C. MeOH (∼20 mL) was added until the blue color disappeared. Water (60 mL) was then gradually added and the reaction allowed warm to room temperature overnight. Extraction was performed with diethyl ether (3 × 80 mL) and the combined organic extracts were washed with water (4 × 50 mL), dried over $MgSO_4$ and the solvent removed under reduced pressure. Thus, the product (19.89 g, 94%) was obtained as a colorless oil.

Dihalides

See Halides.

Diols

31 From Alkenes

Alkenes are oxidized to diols with $KMnO_4$ or with OsO_4.

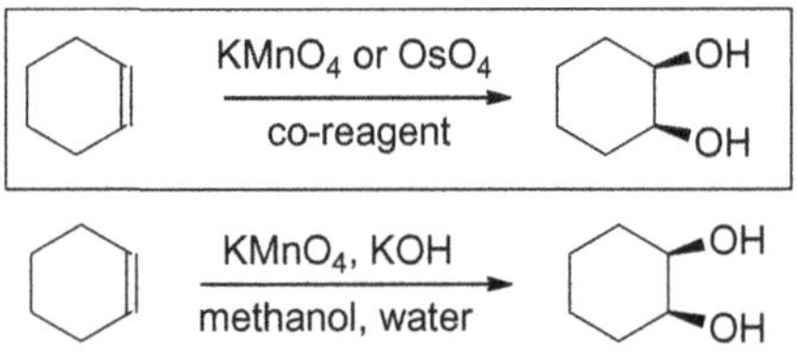

Tetrahedron Lett. **1981**, 22, *2051.*

Potassium permanganate (0.0492 mol) was dissolved in water (450 mL) and this solution was added concurrently with methanol (50 mL) over a period of 1 h to a rapidly stirred slurry of cyclohexene (0.0492 mol) at 20°C, dilute sodium or potassium hydroxide (180 mL) and methanol (20 mL). After treatment with SO_2 to remove manganese dioxide, and aqueous sodium bicarbonate (∼pH 8), the glycol was extracted continuously with dichloromethane for 2−3 d. This solution was concentrated to ~6 mL/g glycol and was then allowed to crystallize in an ice chest, giving a good yield of glycol. A second crop of lower purity was obtained by concentrating the solution.

$$\text{OsO}_4, \text{NMO} \atop \text{acetone, water}$$

Org. Synth. **1978**, *58, 43.*

A 250 mL, three-necked round-bottom flask, with a magnetic stir bar and N_2 inlet is charged with N-methylmorpholine N-oxide mono-hydrate (NMO, 0.1097 mole), water (40 mL) and acetone (20 mL). To this solution is added OsO_4 (0.27 mmol) and cyclohexene (0.1 mol). This two-phase solution is stirred vigorously under N_2 at rt. The reaction is slightly exothermic and is maintained at rt with a water bath. During the overnight stirring period, the reaction becomes homogenous and light brown in color. After 18 h, TLC shows the reaction to be complete. Sodium hydrosulfite (0.5 g) and magnesol (a synthetic magnesium silicate, 5 g) slurried in water (20 mL) are added and the slurry is stirred for 10 minutes and the reaction mixture is filtered through a pad of Celite (5 g) on a 150 mL sintered glass funnel. The Celite cake was washed with acetone (3 × 15 mL), the filtrate, combined with the acetone wash is neutralized to pH 7 with 12 N sulfuric acid (6.4 mL). The acetone is evaporated under vacuum on a rotary evaporator. The pH of the resulting aqueous solution is adjusted to pH 2 with 12 N sulfuric acid (2.3 mL) and the cis diol is separated from N-methylmorpholine hydrosulfate by extraction with portions of n-butanol (5 × 45 mL). The combined butanol extracts are extracted once with 25% NaCl solution (25 mL) and the aqueous phase is backwashed with butanol (50 mL). The butanol extracts are evaporated under vacuum, giving 12.1 g of white solid. The *cis* diol is separated from a small amount of insoluble material by boiling the solvent with 200, 80 and 20 mL portions of diisopropyl ether, decanting the solvent each time. The combined ether fractions are evaporated to ~50 mL under vacuum and crystallized white plates precipitate. The mixture is cooled to ~15°C, the crystals are filtered, washed with 2 × 10 mL portions of cold diisopropyl ether and dried, yielding 10.18–10.31 g of *cis* diol.

32 From Epoxides
Epoxides react with water and acid to give diols or with water and alcohols to give hydroxy ethers.

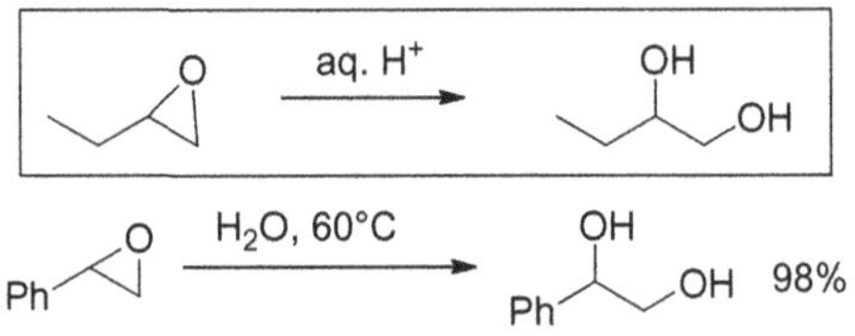

J. Org. Chem. **2008**, *73, 2270–2274.*

A suspension of styrene oxide (1 mmol, 120 mg) in distilled water (6 mL) in a 10 mL flask with a condenser was stirred at 60°C and monitored by TLC. After completion, the mixture was extracted with EtOAc, washed with brine, dried over $MgSO_4$, and then concentrated to give the crude product. Purification by silica gel flash column chromatography provided the diol product (135 mg, 98%).

Enamines

33 From Ketones or Aldehydes

Aldehydes and ketones react with secondary amines to form enamines.

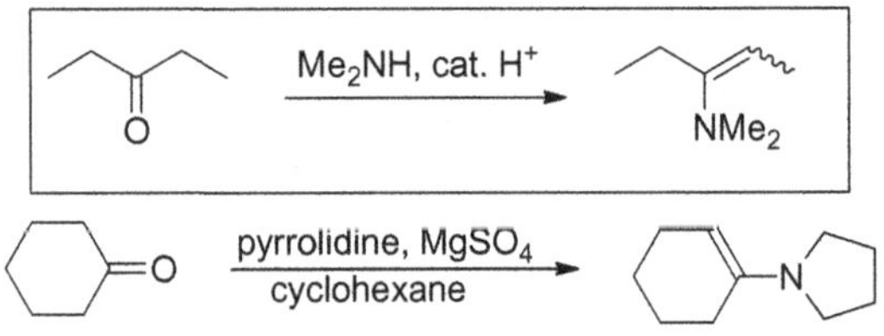

Tetrahedron **2009**, *65, 512–517.*

Cyclohexanone (0.2 mol) was dissolved in cyclohexane (250 mL) and then anhydrous magnesium sulfate (120 g) was added in one portion under a nitrogen atmosphere. The mixture was cooled to 0°C with an ice bath and pyrrolidine (1.00 mol) was added dropwise via syringe over a 0.5 h period. After the reaction mixture had been stirred an additional 30 min at 0°C, the ice bath was removed and the mixture was stirred overnight at room temperature. Magnesium sulfate was removed by filtration and rinsed thoroughly with dry cyclohexane (3×50 mL). The combined filtrate and rinsings were concentrated under reduced pressure to give crude product as an orange oil that was distilled under reduced pressure to give the product as a colorless oil in 93% yield.

Epoxides

34 From Alkenes

Alkenes are oxidized to epoxides with peroxyacids or with dimethyldioxirane. Conjugated ketones or esters react with *tert*-butylhydroperoxide and base to give epoxy-ketones or epoxy esters.

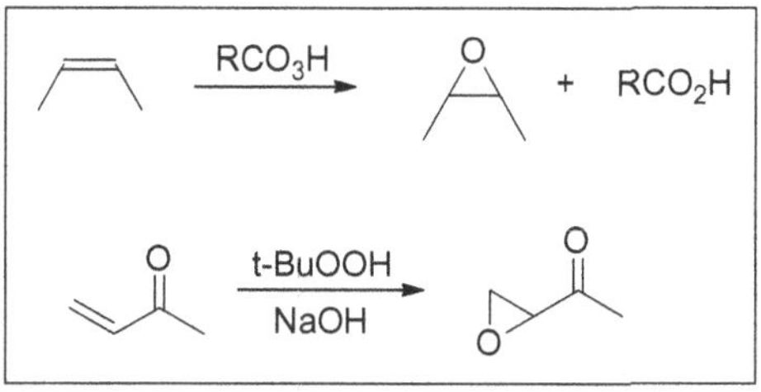

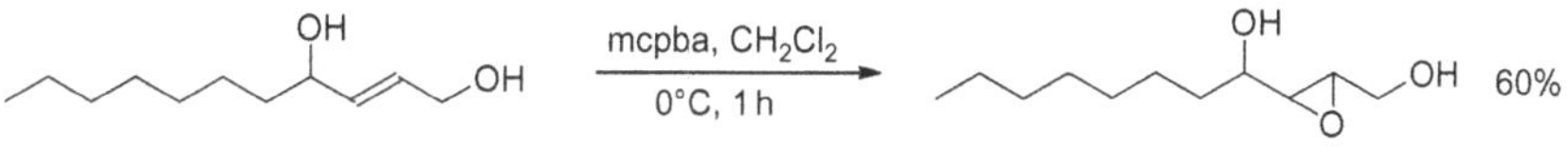

J. Am. Chem. Soc. **1979**, 101, 987–994.

A solution of (Z)-undec-2-ene-1,4-diol (5.0 g, 26 mmol) in CH_2Cl_2 (100 mL) was cooled to 0°C and 85% *m*-chloroperoxybenzoic acid (mcpba, 5.27 g, 26 mmol) was added in one portion. The solution was stirred for 1 h at 25°C and then quenched with saturated $Na_2S_2O_3$ until a negative starch-iodine test was achieved. The solution was washed with 10% Na_2CO_3 to a pH of 9. The organic layer was dried, concentrated, and chromatographed (Et_2O/C_6H_6) to yield 2,3-epoxyundecane-1,4-diol, 3.0 g (60%). Note that benzene is a cancer suspect agent.

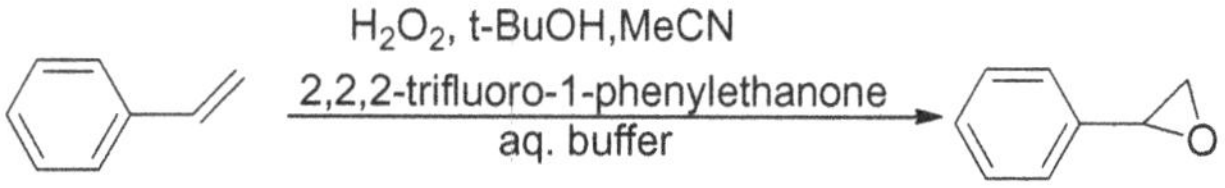

J. Org. Chem. **2014**, 79, 4270–4276.

Styrene (1 mmol) was placed into a round-bottom flask followed by 2,2,2-trifluoro-1-phenylethanone   (0.05 mmol).   *tert*-Butyl   alcohol (1.5 mL), aq buffer solution (1.5 mL, 0.6 M K_2CO_3, 4×10^{-5} M EDTA tetrasodium salt), acetonitrile (2.00 mmol) and 30% aq. H_2O_2 (2.00 mmol) were added consecutively. The reaction mixture was stirred for 1 h at rt. The crude product was purified by flash column chromatography.

35 From Halohydrins

Alkenes are converted to halohydrins and subsequent treatment with a base gives the alkoxide. An internal Williamson ether synthesis generates the epoxide.

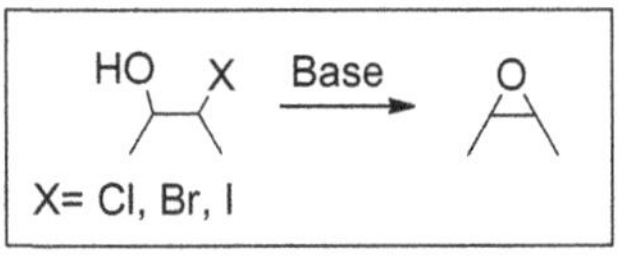

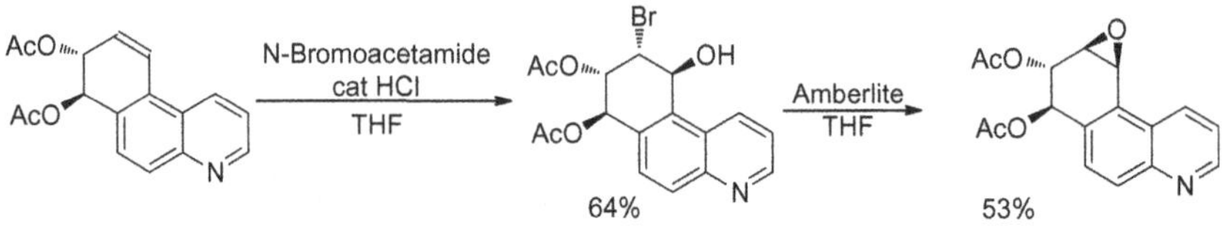

J. Org. Chem. **1986**, 51, 3407–3412.

To a stirred solution of **A** (650 mg, 2.18 mmol) and NBA (*N*-bromoacetamide, 310 mg, 2.24 mmol) in THF (50 mL) and H_2O (10 mL) was added a drop of concentrated HCl at 0°C under argon. The progress of the reaction was monitored by silica gel TLC (5% MeOH in CH_2Cl_2). After the reaction was complete (3 h), H_2O (10 mL) and EtOAc (150 mL) were added. The organic layer was separated and washed with H_2O (10 mL), dried (Na_2SO_4), and distilled under reduced pressure to obtain a crude product, which was triturated with ether and filtered to give 467 mg of pure **B**. The filtrate was concentrated and purified on two silica gel plates (20 cm × 20 cm × 1 mm) using 5% MeOH in CH_2Cl_2 to obtain an additional 82 mg of pure **B**: total yield, 549 mg (64%). See reaction 07.

A solution of **B** (60 mg, 0.15 mmol) in dry THF (50 mL) was stirred with Amberlite (1 g) at 25°C under argon for 8 h. The reaction mixture was filtered and the residue was washed with dry THF (4 × 5 mL). The combined THF filtrates were concentrated under reduced pressure to get a crude product which was chromatographed on a silica gel plate (20 cm × 20 cm × 1 mm) using 5% MeOH in CH_2Cl_2 as an eluent to obtain pure **C** (25 mg, 53%) as a crystalline solid.

Esters

36 From Carboxylic Acids and Acids Derivatives

Carboxylic acids react with alcohols to give an ester. Acid chlorides, acid anhydrides, and also esters react with alcohols to give an ester.

J. Agric. Food. Chem. **2004**, 52, *7480–7483.*

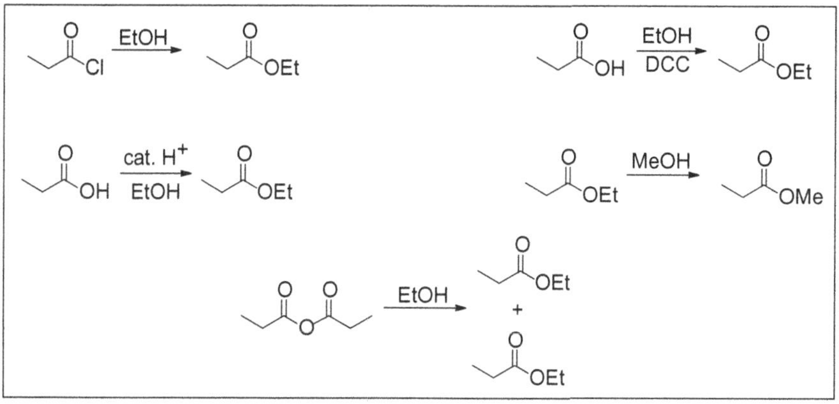

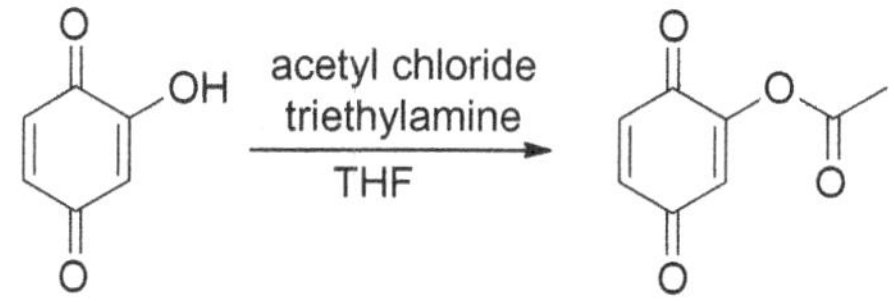

To a solution of p-hydroxyquinone (53.0 mmol) and triethylamine (106 mmol) in THF (50 mL) at −30°C, acetyl chloride (53.0 mmol) was added dropwise over 20 min. After the mixture was stirred for 2 h, the reaction mixture was concentrated *in vacuo* and the residue was dissolved in ethyl acetate. The resulting solution was washed with water and brine, dried with magnesium sulfate and concentrated. The oily residue was purified by column chromatography to give the product in 52% yield.

Synth. Commun. **2013**, 43, *34–43.*

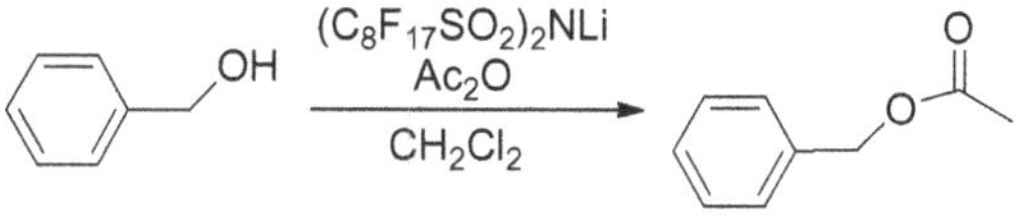

Benzyl alcohol (1.0 mmol), Ac_2O (2 mmol) and $(C_8F_{17}SO_2)_2NLi$ (0.05 mmol) were successively added to a reaction tube. The resulting solution was stirred under 30°C until TLC showed the reaction was completed. Dichloromethane (DCM, 10 mL) was added and the precipitate of $(C_8F_{17}SO_2)_2NLi$ was separated through centrifugation to recover the catalyst (49 mg). The layer of DCM solution was concentrated to give a crude product, which was purified by flash column chromatography on silica gel (pet ether/ethyl acetate = 8/1) to provide the desired product in 98% yield.

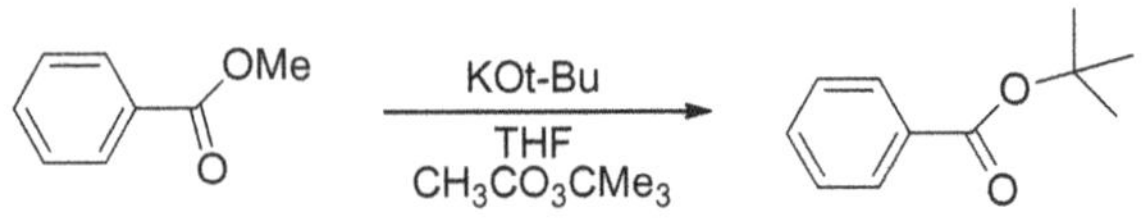

J. Am. Chem. Soc. **1997**, 119, 5075–5076.

Methyl benzoate (25 mmol) and *tert*-butyl acetate (50 mmol) were combined in a 50 mL Schlenk flask. To this mixture of liquids was added a solution of potassium *tert*-butoxide in THF (1 mol%, 0.25 mmol) *via* syringe. The reaction vessel was stirred under dynamic aspirator vacuum at 45°C for 10 min to remove methyl acetate. An additional 1 mol% of catalyst was added every 5 min until the total catalyst concentration reached 5 mol%. Analysis of the reaction mixture by GC indicate that the reaction had proceeded to >99% conversion. The catalyst was removed by passing the reaction mixture through a plug of silica gel and eluting with diethyl ether. The eluent was then evaporated *in vacuo* to yield 4.4 g (98%) of a colorless liquid.

37 From Ketones

Ketones react with peroxyacids to give esters.

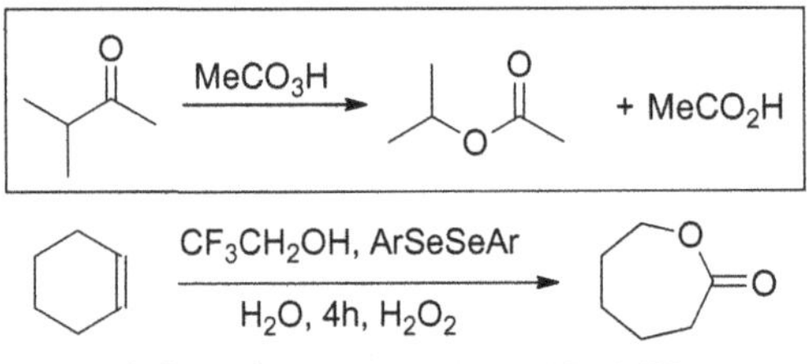

J. Org. Chem. **2001**, 66, 2429–2433.

Hydrogen peroxide (20 mmol) was added to a solution of bis(3,5-trifluoromethylphenyl) diselenide (0.1 mmol) in CF_3CH_2OH (10 mL). After 10 min, cyclohexanone (10 mmol) was added. The mixture was stirred for 4 h and water was added to give a volume of 100 mL. The aqueous mixture was extracted with ether (4 × 20 mL). The organic layer was washed with 10% $NaHSO_3$ (20 mL), 10% $NaHCO_3$ (20 mL) and water (20 mL) and then dried over sodium sulfate. The solvent was removed *in vacuo* and the residue was purified by bulb-to-bulb distillation to give 56% yield of the product.

Ethers

38 From Alcohols

Alcohols react with a strong base to generate an alkoxide, and S_N2 reaction with an alkyl halide gives the ether (Williamson Ether Synthesis).

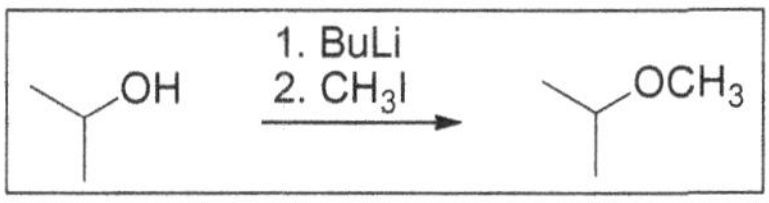

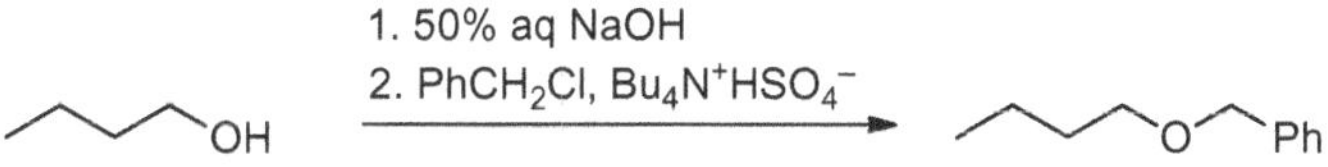

J. Chem. Educ. **1980**, 57, 822.

In a round-bottom flask, a 50% solution of sodium hydroxide was prepared by dissolving 20 g of NaOH (0.5 mole) in 20 mL of water. After cooling to room temperature, 7.4 g (9.2 ml, 0.1 mol) of butan-1-ol was added. A thick, white slurry was obtained. At this time, 15.2 g (13.8 ml, 0.12 mole) of benzyl chloride (CAUTION: Lachrymator) was added, and also 1.7 g (0.005 mole) of tetrabutylammonium hydrogen sulfate. A reflux condenser was attached and the reaction stirred on a water bath at 75°C for 15 min. After stirring, the reaction was extracted with diethyl ether. The ether extract was dried over anhydrous calcium chloride and the diethyl ether removed by simple distillation. When all the diethyl ether had been distilled, the condenser water was disconnected and the residue heated strongly to distill the benzyl butyl ether through an air-cooled condenser. Benzyl butyl ether boiled at 200°C. The product was obtained in good yield (50–90%, average 67%).

Halides (Alkyl)

39 From Alcohols

Alcohols react with HBr or HCl to form alkyl bromides or alkyl chlorides; alcohols react with sulfur or phosphorus halides to form alkyl halides.

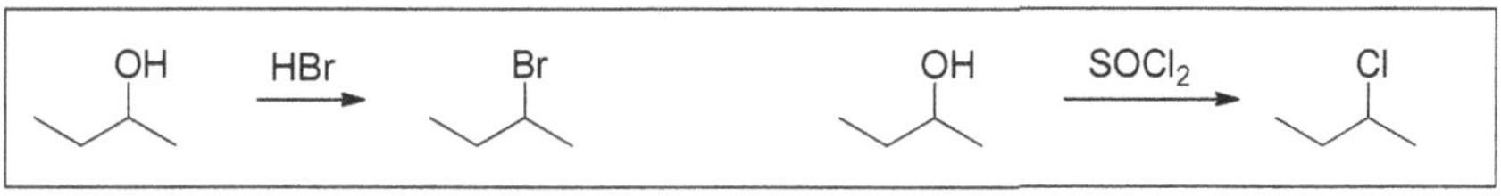

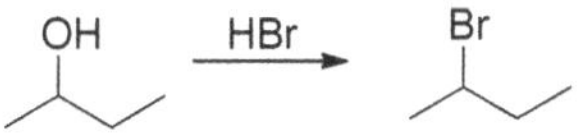

J. Am. Chem. Soc. **1962**, 84, 4660–4661.

Hydrogen bromide was bubbled through (+)-butan-2-ol (31.05 g, $[\alpha]^{25}_D + 10.6°$), cooled in an ice bath until the theoretical weight had been taken up. The product was sealed in several tubes and heated to 110°C for 3 h. After the tubes were cooled in ice, they were opened and the product separated from the heavy acid layer. The upper layer was extracted with a concentrated solution of calcium chloride to remove unreacted alcohol and then dried over calcium chloride. The bromide was distilled to give a yield of 42.7 g (73%) $[\alpha]^{25}_D -16.65°$.

40 From Alkanes and Aryls

Alkanes are generally unreactive, but they react with bromine or chloride under radical conditions to give alkyl halides. Benzene reacts with bromine, chlorine, nitric acid/sulfuric acid or SO_3/sulfuric acid in the presence of a Lewis acid catalyst via electrophilic aromatic substitution.

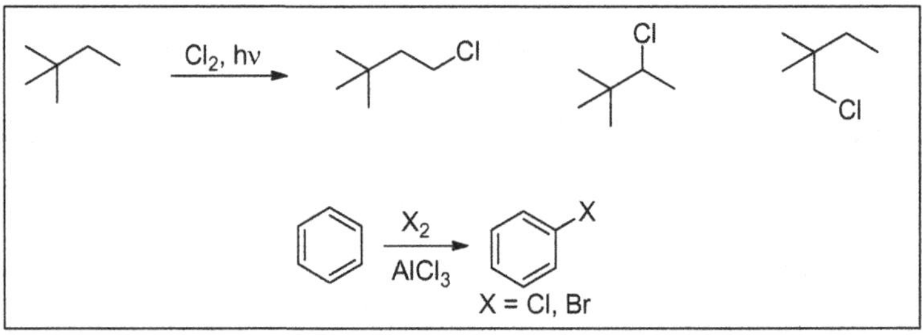

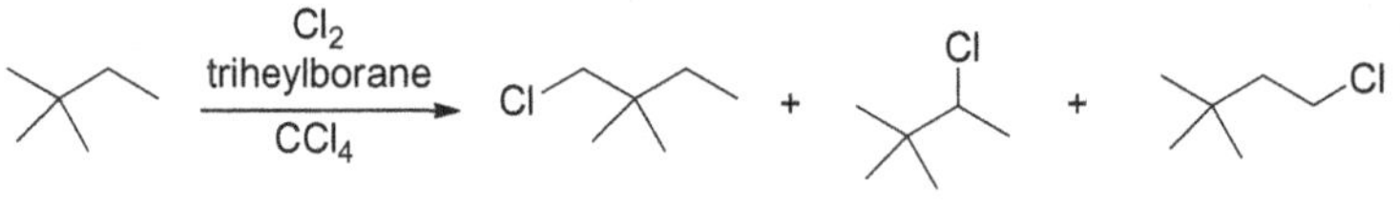

Chem. Lett. **1984**, 13, 195–198.

2,3-Dimethylbutane (60 mmol) and trihexylborane (0.3 mol) were placed in a 50 mL round-bottom flask, protected from light by a black plastic bag and flushed with argon. To the stirred solution, chlorine (6 mmol) in carbon tetrachloride (6 mL) was slowly added at 20°C after 2 h, of the reaction under the weak stream of argon, the reaction mixture was neutralized with aqueous sodium hydroxide and washed several times with NaCl-saturated water.

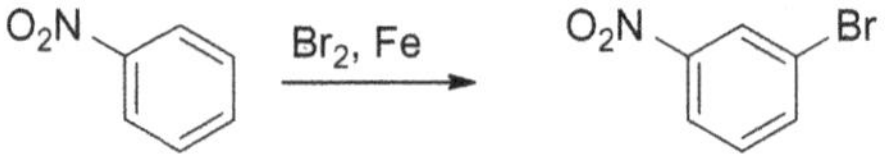

Org. Synth. **1928**, 8, 46.

Freshly distilled dry nitrobenzene (2.2 mol) was added to a 3 L three-necked round-bottom flask, fitted with an efficient reflux condenser bearing an outlet tube held above a surface of water, a 100 mL separatory funnel, and a mercury-sealed mechanical stirrer. The joints in the apparatus were made of asbestos paper covered with water glass. The flask was heated in an oil bath maintained at 135−145°C and iron powder (26 g) and dry bromine (3.5 mol) was added in the following manner: Iron powder (8 g) ("ferrum reduction") was added through the side neck to the stirred nitrobenzene. From the separatory funnel, bromine (60 mL) was added at such a rate that the bromine vapors did not traverse the condenser. This addition required about 1 h and the mixture was stirred and heated for another hour before the addition of a second portion of iron and bromine. Two portions, each iron powder (8 g) and bromine (60 mL) were added under the same conditions as the first addition and the mixture was stirred and heated for 1 h between the completion of one addition and the beginning of another. The evolution of hydrogen bromide slackened considerably toward the last of the heating and there was practically no more bromine vapor in the condenser. A final addition of iron powder (2 g) was made and the heating continued for 1 h longer. The reaction product, which was a dark reddish-brown liquid, was poured or siphoned into water (1.5 L) to which a saturated solution of sodium bisulfate (50 mL) had been added. The mixture was distilled with steam and the first portion of the distillate collected separately to remove a small amount of unchanged nitrobenzene. It was necessary to collect about 12 L of distillate in order to obtain all of the m-bromonitrobenzene. The yellow crystalline solid was filtered with suction and pressed well on the funnel to remove water and traces of nitrobenzene. The yield of crude product varied from 270 to 340 g (60−75%).

41 From Alkenes

Alkenes react with Brønsted-Lowry acids, H—X, to give alkyl halides. Conjugated dienes undergo primarily 1,4-addition with HCl or HBr. Alkenes react with dihalogens, X_2, to give alkyl dihalides. Conjugated dienes undergo primarily 1,4-addition with Br_2 at >50°C and 1,2-addition at <25°C.

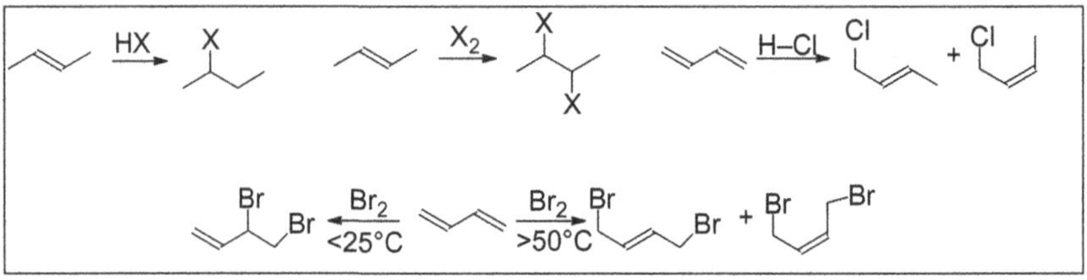

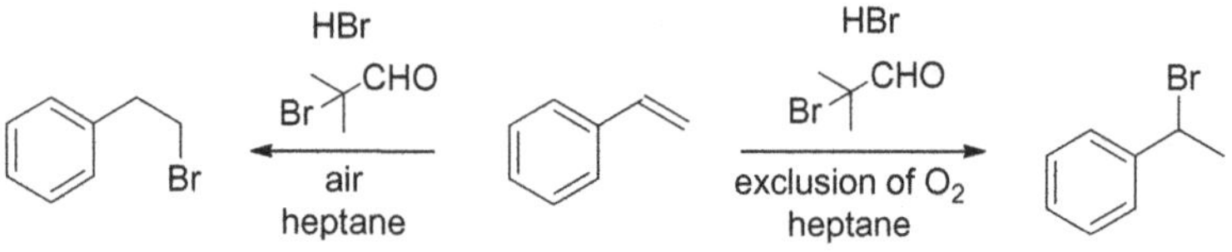

J. Org. Chem. **1995**, 60, 1315–1318.

A 100 mL 3-necked flask (painted black) was equipped with a magnetic stirrer, gas inlet, thermometer and HBr trap with 40 mL solvent and the required amount of 2-bromo-2-methyl propanal or other additive. Gaseous HBr was added after drying over P_2O_5. A 1 mL sample was taken to measure the HBr concentration and the reaction was initiated by addition of styrene. After 30 min, the styrene conversion was essentially quantitative. A 5 mL aliquot of reaction mixture was washed 3x with water, 5% $NaHCO_3$ and water, dried over Na_2SO_4 and analyzed by GLC.

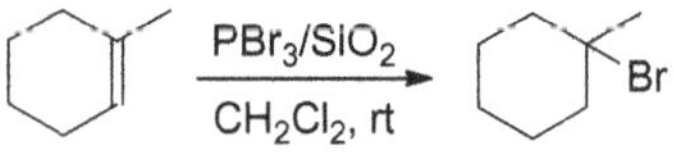

J. Braz. Chem. Soc. **2001**, 12, 685–687.

To a stirred suspension of the alkene (10 mmol) and SiO_2 (5 g) in CH_2Cl_2 (25 mL), a solution of PBr_3 (4 mmol in CH_2Cl_2 (10 mL) was added for 10 min at rt. After the addition was complete, the suspension was stirred for several minutes and then filtered. The SiO_2 was washed with CH_2Cl_2 (15 mL), the combined liquid was washed with 10% $NaHCO_3$ (until no more gas liberated), brine (2x) and the organic extract was dried (Na_2SO_4). The solvent was evaporated on a rotary evaporator under reduced pressure to give pure bromide.

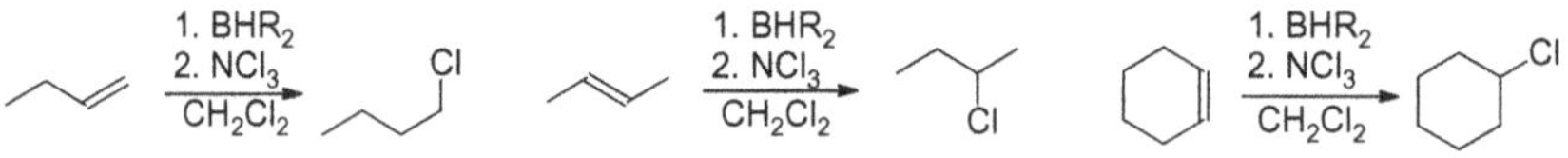

Tetrahedron **1998**, 44, 2785–2792.

A 50 mL flask equipped with magnetic stirrer, septum inlet and pressure equalizing addition funnel was flushed with N_2 and charged with tri-n-butyl borane (2 mmol) and CH_2Cl_2 (2 mL) and cooled to 0°C. Then, NCl_3 (6 mmol) in CH_2Cl_2 was added and the reaction mixture irradiated with a Sears 275 W sunlamp. After 1 h, the reaction mixture was completely decolorized and contained a white precipitate.

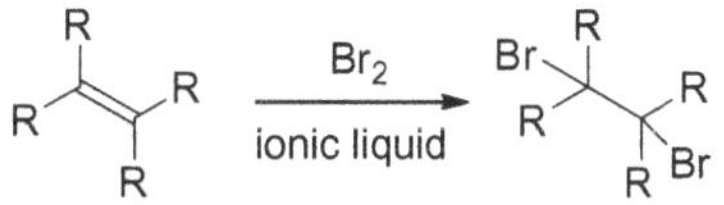

Organic Lett. **2001**, *1061–1063.*

To a round-bottom flask equipped with a magnetic stir flea (a stir bar), the ionic liquid (2 mL) and Br_2 (0.6 mmol) were added. To the Br_2 solution, maintained in a dark room at room temperature, an equimolar amount of the unsaturated compound was added under stirring. A water bath was used in order to avoid temperature increase during alkene or alkyne addition. Products were then extracted at the end of the reaction by three subsequent additions of ether, followed by the decanting off of the ethereal solution of the product. The combined extracts were concentrated on rotary evaporator and the products were analyzed by ^1H NMR.

42 From Alkyl Halides

Alkyl halides undergo substitution reactions with various nucleophiles. Other halides, ethers, nitriles, and alkynes can be prepared.

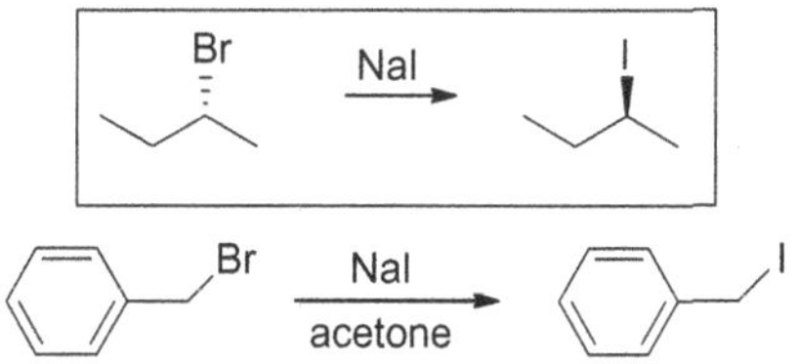

J. Org. Chem. **2009**, *74, 4177–4187.*

Benzyl bromide (60 mmol) was added to a solution of sodium iodide (120 mmol) in acetone (80 mL). The mixture was stirred for 24 h in the dark at room temperature, then quenched with water (50 mL) and extracted with ether (2 × 100 mL). The combined organic layers were dried ($MgSO_4$) filtered, and concentrated under reduced pressure to afford the pure product, quantitatively, as a yellow oil.

Halides (Alkenyl)

43 From Alkynes

Alkynes react with Brønsted-Lowry acids, H—X, to give vinyl halides. Alkynes react with dihalogens, X_2, to give vinyl dihalides.

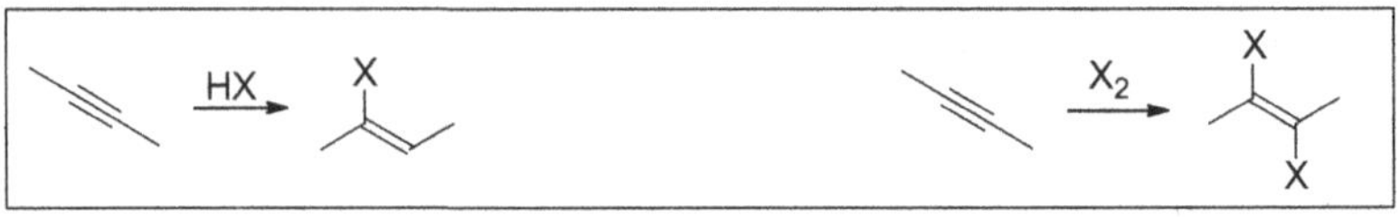

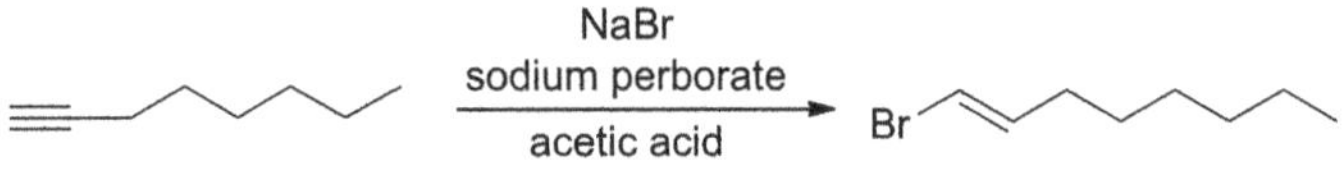

Synth. Commun. **1998**, 28, 3807–3809.

Sodium bromide (10 mmol) was added to the mixture of sodium perborate (10 mmol) and oct-1-yne (5.0 mmol) in glacial acetic acid (25 mL) and stirred for 1.5 h. The mixture was then extracted into carbon tetrachloride (10 mL), washed with water (10 mL) and dried over anhydrous magnesium sulfate. The yield was 89.7% as measured by ^{1}H NMR using 1,4-dioxane as an internal standard.

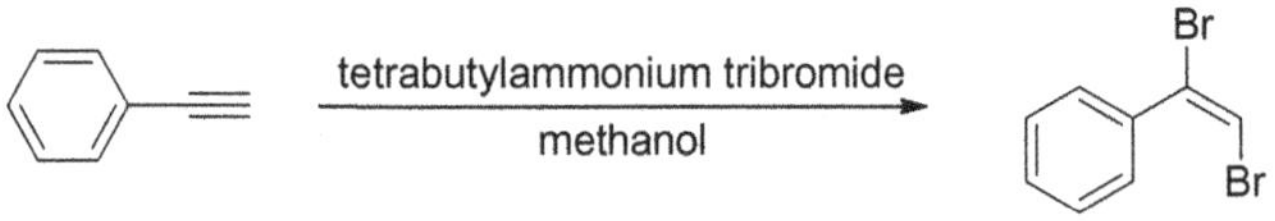

Synth. Commun. **1997**, 27, 2865–2876.

Reactions were generally performed at room temperature. The alkyne (2×10^{-3} mol) and TBABr$_3$ (2×10^{-3} mol) were dissolved in solvent (100 mL). The reactions were carried out either using mechanical stirring or ultrasound. The initially red-colored solution changed gradually to yellowish and finally became colorless at which point the reaction was complete, as confirmed by TLC. The solvent was evaporated and the residue, dissolved in diethyl ether, was successively washed with an aqueous 5% sodium thiosulfate solution and water. After evaporation of the solvent, a colorless oil was obtained. The dibrominated product, the (E)-dibromoalkene) was recrystallized from a few drops of ethanol.

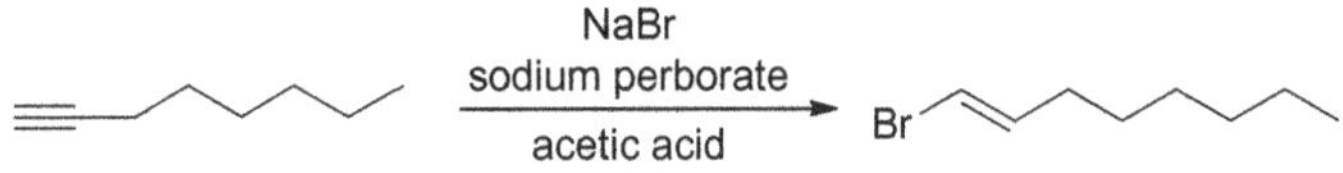

Synth. Commun. **1998**, 28, 3807–3809.

Sodium bromide (10.0 mmol) was added to the mixture of sodium perborate (10 mmol) and 1-octyne (5.0 mmol) in glacial acetic acid (25 mL) and stirred for 1.5 h. The mixture was then extracted into CCl_4 (10 mL), washed with water (10 mL) and dried over anhydrous $MgSO_4$. The yield was 89.7% as measured by ^1H NMR.

Imines
44 From Ketones or Aldehydes
Aldehydes and ketones react with primary amines to form imines.

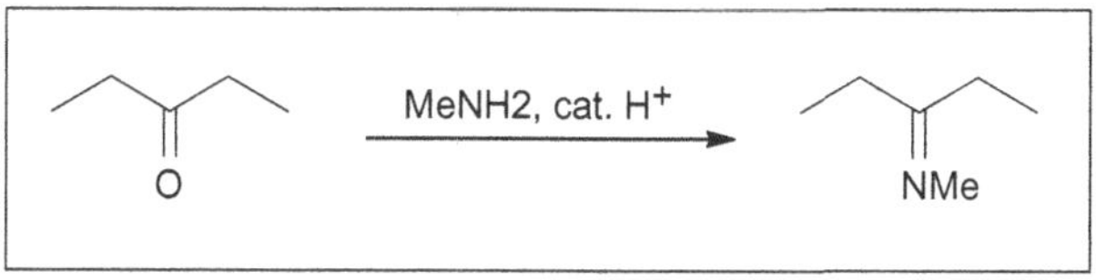

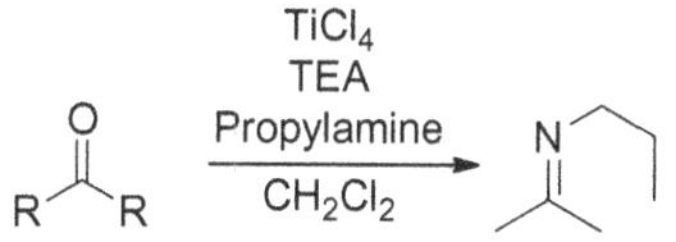

J. Heterocyclic Chem. **2000**, 37, 1309–1314.

A solution of titanium (IV) chloride (30 mmol) in dichloromethane (40 mL) was added slowly, under stirring to a cooled (with an ice bath) solution of propylamine (60 mmol) and trimethylamine (110 mmol) in dichloromethane (60 mL). Subsequently, the ketone (50 mmol) was added at once. After stirring the mixture for 24 h at room temperature the solvent was removed on the rotary evaporator. The residue was crushed with a spatula. Then diethyl ether (200 mL) was added and the resulting mixture stirred vigorously until the residue was ground to a fine powder. Subsequently, the powder was removed *in vacuo* and washed with diethyl ether (200 mL). Evaporation of ether yielded the crude imine, which was used for the synthesis of pyrazoles without prior purification.

Ketals
See Acetals.

Ketones or Aldehydes. See Aldehydes
45 From Alkynes
Alkynes react with weak acids such as water and alcohols, using a catalytic amount of a strong acid to give ketones or vinyl ethers, respectively. Alkynes react with mercuric compounds and water to give ketones or

aldehydes; internal alkynes react with boranes and then hydrogen perox-ide/NaOH to give ketones or aldehydes.

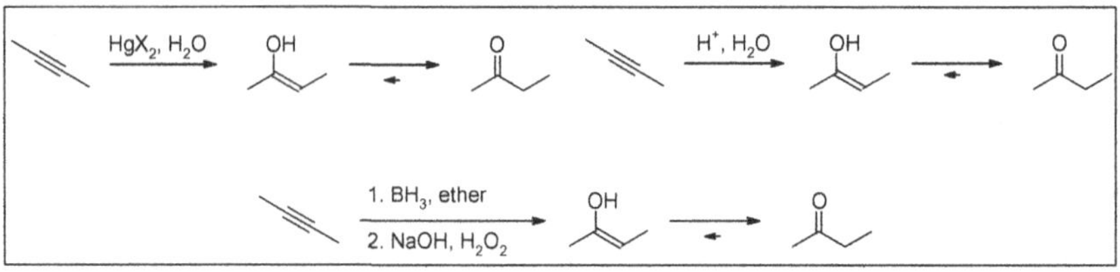

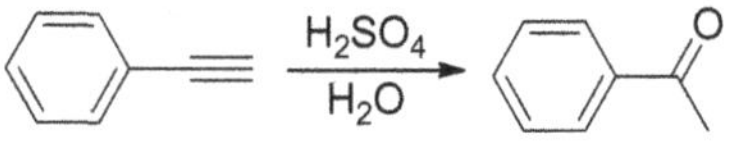

ACS Catal. **2011**, *1, 116–119.*

A bis(*N*-methylpyrrolidinium) dihydrogen sulfate ionic liquid [1,1′-(hexane-1,6-diyl)bis(1-methylpyrrolidin-1-ium), dihydrogen sulfate 7 mol], sulfuric acid (0.430 mL, 8 mmol), water (2 mmol) and phenyla-cetylene (1 mmol) were added to a 25 mL round-bottom flask. The reaction mixture was vigorously stirred at 40°C for 30 min and then extracted with pentane (4 × 10 mL). The collected organic layers were combined and dried over sodium sulfate. The solution was concentrated by rotary evaporation and the yield was determined by GC with an internal standard.

46 From Diols
Diols are oxidatively cleaved to aldehydes and ketones with periodic acid or with lead tetraacetate.

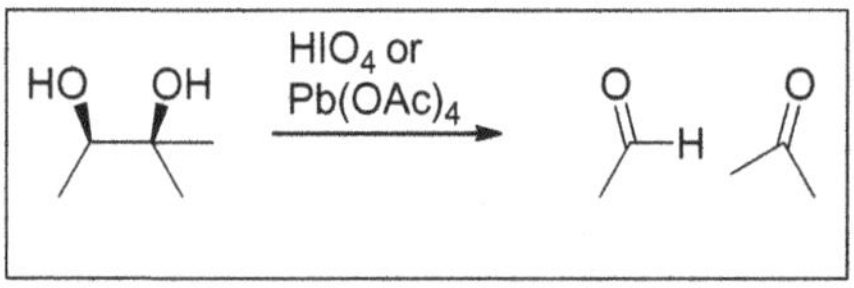

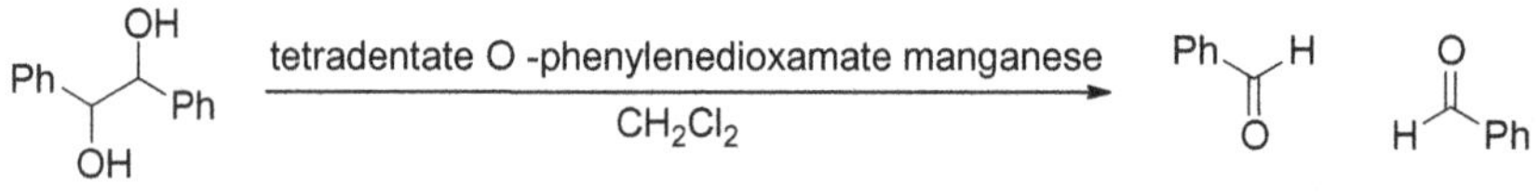

Catalysis Commun. **2008**, *9, 1282–1285.*

The diol (0.27 mmol) is dissolved in dichloromethane (2 mL) and added to a round-bottom flask that contains an ionic liquid (5 mL) and [Mn(salen)(Py)](OAc) (0.014 mmol). The organic solvent is evaporated and the system is bubbled constantly with oxygen at 30°C. One equivalent of the sacrificial aldehyde is added and the addition is repeated after 1 and 2 h. After 3 h, the oxygen bubbling is suspended and heptane (5 mL) was added to the system. The products are extracted and purified by flash chromatography to give the oxidized product.

47 From Ketone-Acids

Heating a β-keto acid leads to decarboxylation to give a ketone.

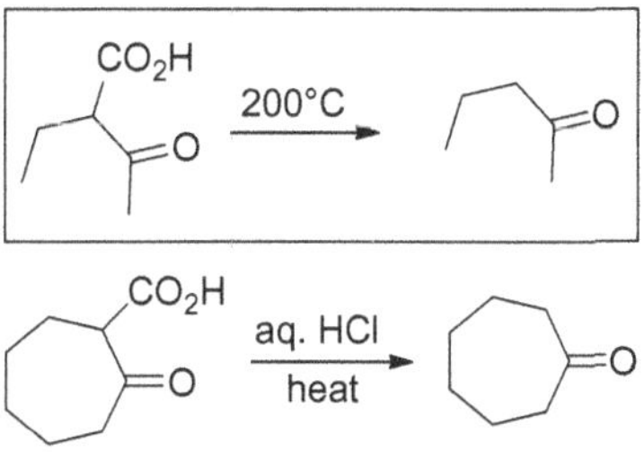

J. Org. Chem. **1958**, 23, 1708–1710.

A dilute solution of hydrochloric acid (about 3 N) was added to 2-oxocycloheptane-1-carboxylic acid. Hydrolysis and decarboxylation were accomplished by heating at reflux overnight. Ethanol was added to the decarboxylation medium to promote solubility of the keto-esters. The unreacted starting material was recovered as diester rather than as an acid. The reaction mixture was extracted with ether, and the combined ether extracts were extracted with a 10% aq solution of sodium bicarbonate. The ethereal solution was distilled and the cycloheptanone was collected.

Nitriles

48 From Amides

Amides (primary) react with P_2O_5 to give a nitrile.

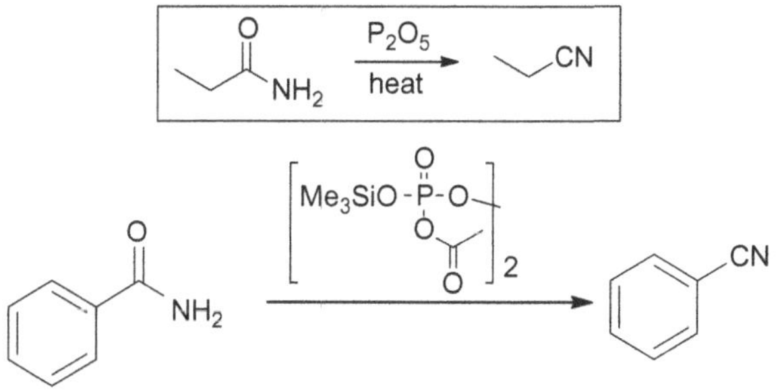

Synth. Commun. **1989**, 19, 1431–1436.

Phosphorus pentoxide (0.0075 mol) was treated at 20°C with a silyl sulfonate (0.015 mol) and the resulting mixture heated under stirring at 50°C for 30 minutes. The complex thus obtained was cooled to room temperature (30°C), treated with the amide (0.01 mol) and then stirred and heated at 70−75°C for 3 h. The reaction mixture was cooled in an ice-bath and quenched with saturated sodium bicarbonate solution under stirring. The product was extracted into benzene or ether and the extract passed through a bed of basic alumina to get rid of highly acidic phosphorus-residues, which otherwise contaminated the product even after distillation. After removal of the solvent the nitriles were purified by Kügelrohr distillation.

Nitro Compounds

49 From Aromatics

Benzene reacts with nitric acid/sulfuric acid in the presence of a Lewis acid catalyst via electrophilic aromatic substitution (S_EAr reactions).

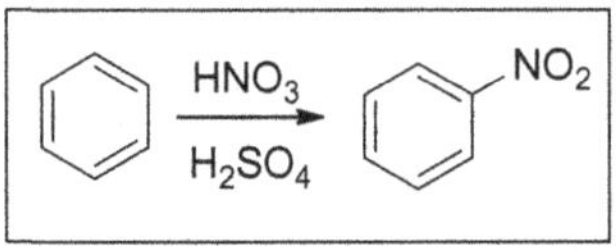

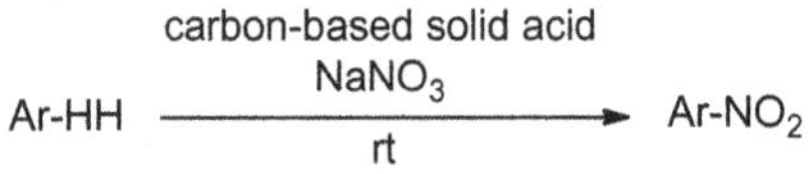

$$Ar\text{-}HH \xrightarrow[\text{rt}]{\substack{\text{carbon-based solid acid}\\ \text{NaNO}_3}} Ar\text{-}NO_2$$

Chinese Chem. Lett. **2007**, 18, 1064−1066.

A mixture of substrate (1 mmol), sodium nitrate (1 mmol) and a wet carbon-based solid acid [70% (w/w), 0.7 g] was pulverized in a mortar at room temperature for the specified time. The progress of the reaction was monitored by TLC method after completion of the reaction mixture extracted with dichloromethane (2 × 10 mL). Dichloromethane was finally removed and nitro compounds were obtained in good yield.

Organometallics

50 From Alkyl Halides

Alkyl halides react with magnesium to form organomagnesium reagents (Grignard reagents). Alkyl halides react with lithium metal to form organolithium reagents. Organolithium reagents react with copper salts to form organocuprates.

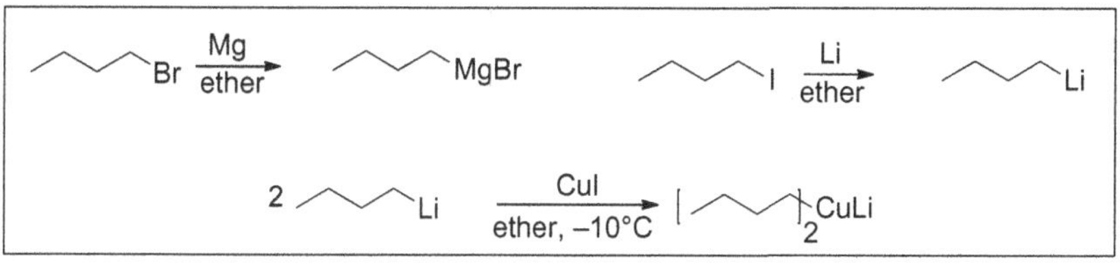

$$CH_3I \quad + \quad Mg \xrightarrow{\text{ether}} CH_3MgI$$

J. Am. Chem. Soc. **1978**, 100, 3163–3166.

Approximately 20 mg of Mg turnings and 10 mL of anhydrous diethyl ether were added to the dry 25-mL round-bottom flask through the septum port while purging with N_2. The mixture was stirred vigorously with a magnetic stirring bar and the ether was brought to reflux temperature. A known quantity (~ 0.5 mmol) of dry CH_3I was injected into the refluxing ether.

After reaction (a variable time of reaction) the resulting slurry was used in further reactions.

Phenols

51 From Aryl Halides

Phenols are generated by S_NAr reactions of aryl halides or by conversion of aromatic amines to aryldiazonium salts, followed by reaction with aqueous acid.

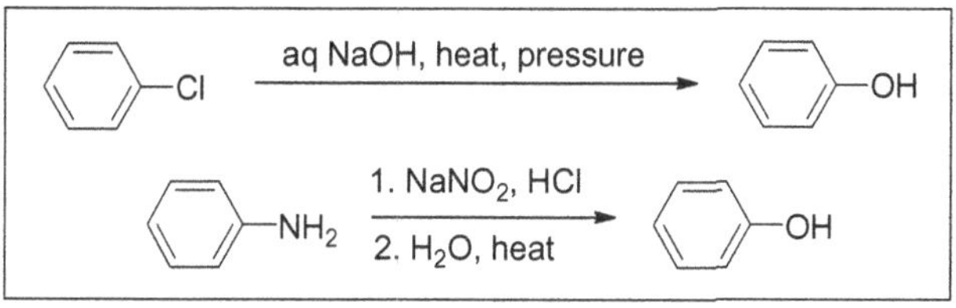

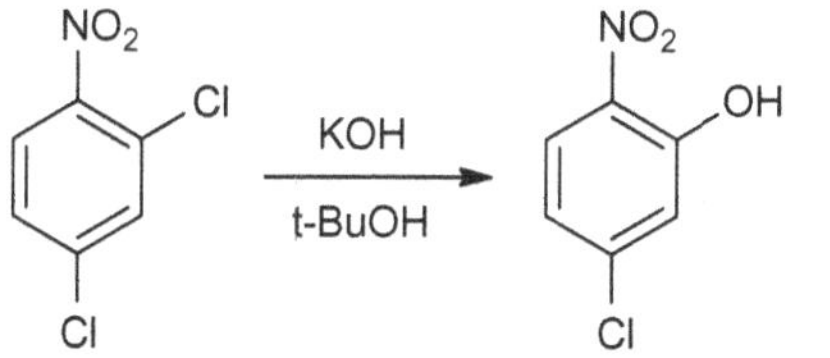

Synth. Commun. **2004**, 34, 2903–2909.

2,4-Dichloro nitrobenzene (0.032 mol) and potassium hydroxide (0.078 mol) were refluxed in *tert*-butanol (25 mL) for 10 h. The reaction mixture was cooled, solvent removed under reduced pressure. The residue was dissolved in water (20 mL) and extracted with dichloro-methane to remove the unreacted product. The aqueous layer was acidi-fied with concentrated hydrochloric acid and extracted with dichloromethane. The organic layer dried over sodium sulfate, concen-trated and on flash chromatography afforded the 2-nitro-5-chloro phenol (4.23 g, 77%).

Phosphonium Salts

52 From Alkyl Halides

Alkyl halides react with phosphines to give phosphonium salts.

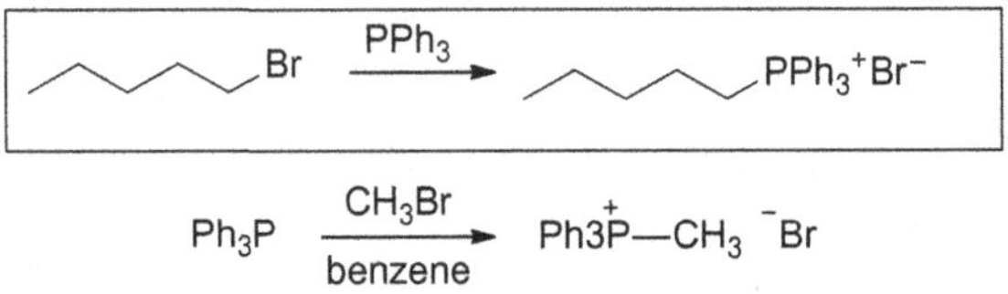

Org. Synth. **1960**, 40, 66.

A solution of triphenylphosphine (0.21 mole) dissolved in dry benzene (45 mL) was placed in a pressure bottle, the bottle cooled in an ice-salt mixture, and methyl bromide (0.29 mole) of previously condensed added. The bottle was sealed and it stood at room temperature for 2 d, and was then reopened. The white solid was collected by means of suction filtra-tion with the aid of hot benzene (500 mL) and then dried in a vacuum oven at 100°C over phosphorus pentoxide. The yield was 74 g. Note. Benzene is a cancer suspect agent.

Sulfonic Acids

53 From Aromatics

Benzene reacts with SO_3/sulfuric acid in the presence of a Lewis acid catalyst or excess SO_3 via electrophilic aromatic substitution.

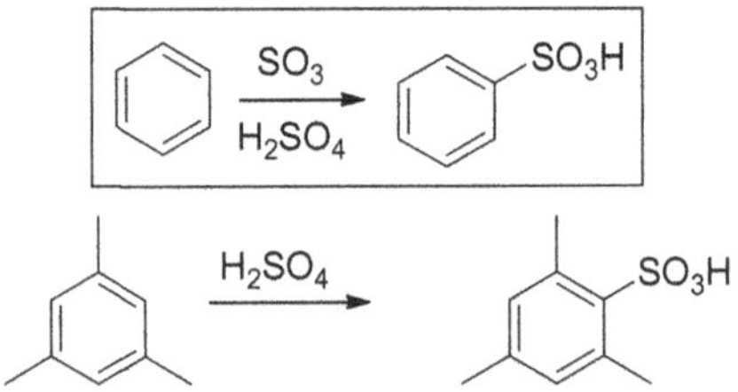

Tetrahedron Lett. **2004**, 45, 6607–6609.

A 25 mL round bottomed flask was charged with silica sulfuric acid (5 mmol) and mesitylene (5 mL) and a magnetic stirrer. The reaction mixture was stirred at 80°C for 30 min, the heterogeneous mixture then was filtered, washed with 10 mL of dichloromethane and the solvent was removed under reduced pressure. The residue was washed with *n*-hexane (2 × 10 mL) and dried in air to produce white solid (0.75 g, 75% yield).

8.B CARBON-CARBON BOND FORMING REACTIONS

Alcohols

54 From Acid Derivatives and Organometallics

Acid chlorides, acid anhydrides, and esters react with Grignard reagents or organolithium reagents to give ketones or alcohols (Grignard reaction, and related reactions).

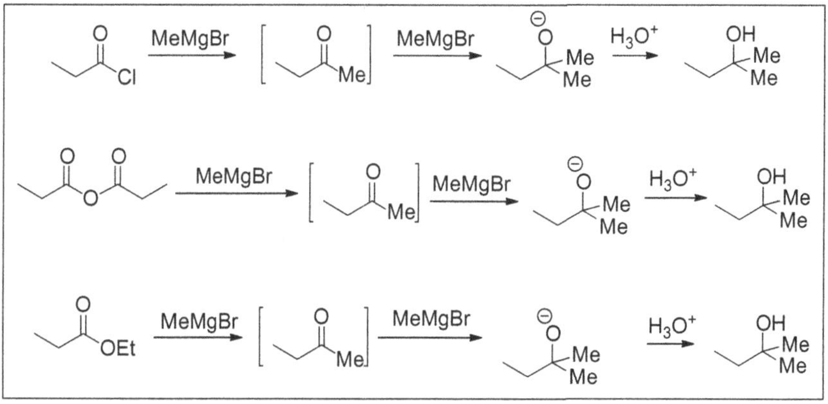

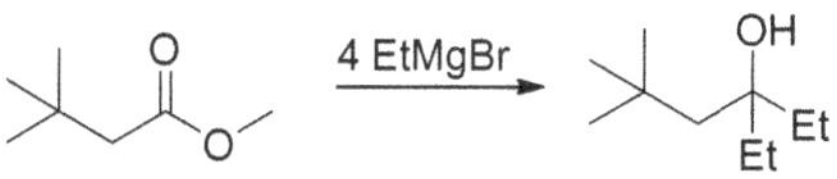

J. Am. Chem. Soc. **1942**, 64, 2966–2968.

Addition of 1 mol of ester to 4 mol of the Grignard reagent was carried out at room temperature; the rate of addition was about 1 mol per h. The products were worked up by shaking with cracked ice, separating the ether and steam distilling the residues. The steam distillates were combined with the respective ether extracts. This method minimized any dehydration of the tertiary alcohols. After stripping off the ether, the products were fractionated through columns of 12–15 theoretical plates to give 108.4 g. or 68.5% of diethylneopentylcarbinol, b.p. 54°C (5 mm).

55 From Epoxides

Epoxides react with Grignard reagents, organolithium reagents, and organocuprates to give the corresponding alcohol.

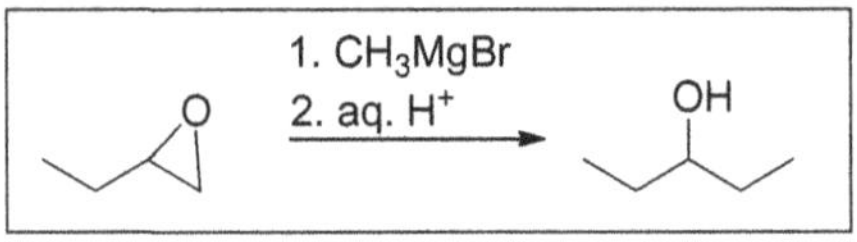

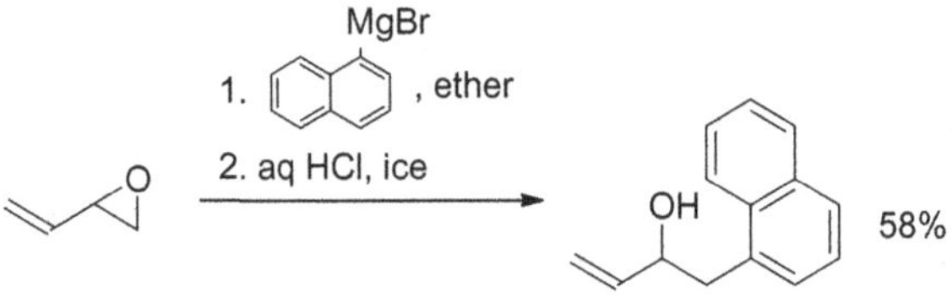

J. Org. Chem. **1950**, 15, 305–316.

1-Naphthylmagnesium bromide was prepared under nitrogen from 6.08 g. (0.25 mmol) of magnesium turnings and 51.8 g. (0.25 mol) of 1-bromonaphthalene in 175 mL of absolute ethyl ether (50 mL). A solution of 17.5 g. (0.25 mol) of 3,4-epoxybut-1-ene in 50 mL of ether was added dropwise with stirring at such a rate as to maintain a vigorous reflux (about 45 min is required). Stirring was continued for 8 h after the reflux and then the reaction mixture was stood at room temperature for 12 h. The solution was hydrolyzed with 100 mL of 2.5 N hydrochloric acid and 200 g of ice, the ether layer separated, and the water layer extracted with ether; the combined ether extracts were washed with water until neutral. After drying the ether layer over potassium carbonate, filtering, and concentrating it at reduced pressure, it was steam-distilled at reduced pressure to remove naphthalene. During this codistillation, which required 2 L of water, the vapor temperature did not exceed 50°C. The residue was extracted with ether and dried twice with fresh potassium carbonate. After distillation of the ether, distillation of the residue gave 27.2–28.6 g (0.137–0.145 mol; 55–58%) of the carbinol.

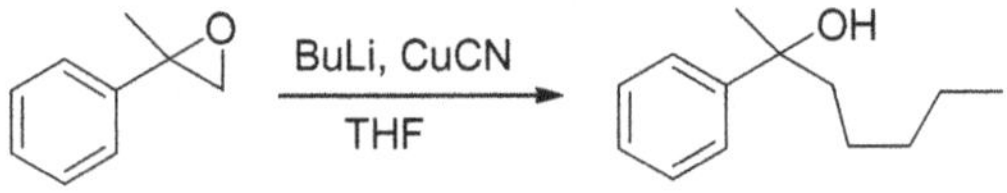

J. Am. Chem. Soc. **1982**, 104, 2305–2307.

Copper cyanide (1.1 mmol) was placed in a two-neck 25 mL round-bottom flask equipped with a magnetic stir bar. The salt was azeotropically dried with toluene (2 × 20 mL) with successive purging with argon. Dry THF (2.0 mL) was added and the slurry cooled to −78°C. *n*-butyllithium (2 mmol) was added dropwise producing a tannish yellow solution, which was warmed to −20°C. Freshly distilled α-methylstyrene oxide (0.75 mmol) was dissolved in 1.0 mL THF, cooled to -20°C and transferred to the cuprate via cannula with subsequent wash with cold (−20°C) THF (0.5 mL). The reaction was stirred at this temperature for 2 h and then quenched with 90% NH_4Cl/10% NH_4OH solution (5 mL). After stirring at room temperature for about 30 min, the solution was transferred to a separatory funnel and 5 mL of saturated aqueous NaCl solution was added. This was then extracted with ether (3 × 7 mL), and the extracts were dried over K_2CO_3. Following filtration, the ether was removed *in vacuo*, yielding a light yellow oil that was chromatographed on silica gel (30% ether/pentane) to afford 0.134 g of a clear oil.

56 From Ketones or Aldehydes and Organometallics

Aldehydes and ketones react with Grignard reagents and organolithium reagents to form a new C—C bond and form an alcohol. Conjugated aldehydes undergo a mixture of 1,2- and 1,4-addition with Grignard reagents, but mostly 1,2-.

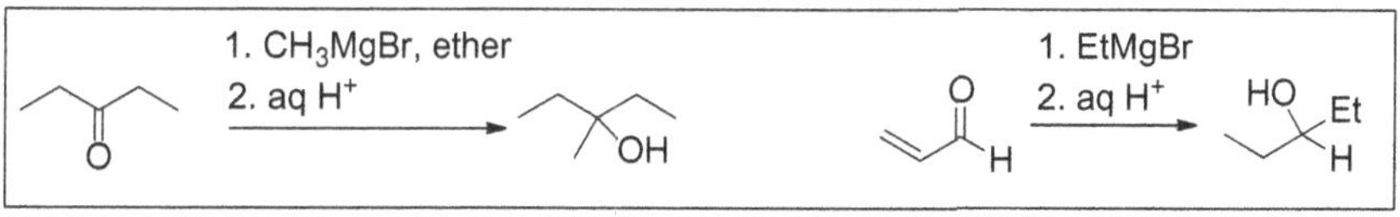

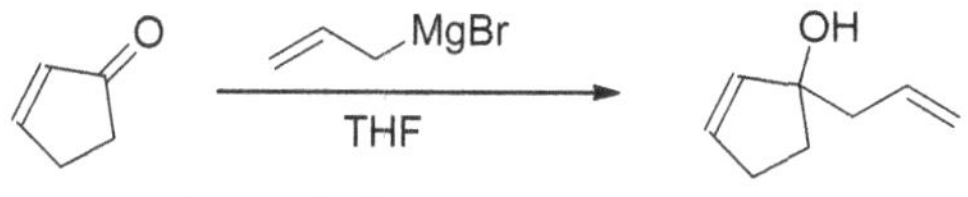

Org. Biomol. Chem. **2008**, 6, 4299–4314.

Allylmagnesium bromide in THF (1.5 equiv) was added dropwise to a stirred solution of cyclopentenone (1.7 mmol) in THF (8 mL) at 0°C under nitrogen. The resulting solution was allowed to warm to rt and stirred at rt for 12 h. Saturated ammonium chloride (5 mL) was added and the layers were separated. The aqueous layer was extracted with ether (3 × 5 mL) and the combined organic layers were dried over magnesium sulfate and evaporated under reduced pressure to give the crude product.

Alcohols-Esters

57 From Esters

Enolate anions of esters react with aldehydes or ketones to give β-hydroxy esters.

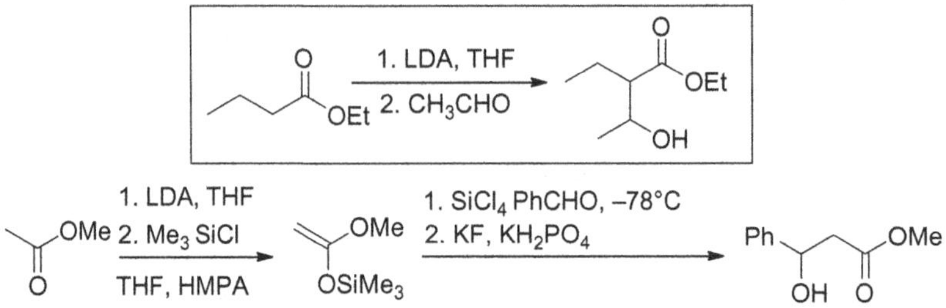

J. Am. Chem. Soc. **2005**, 127, 3774–3789.

To a flame-dried, 100 mL, 3-neck, round-bottom flask was added diisopropylamine (11 mmol) and THF (14 mL). The solution was cooled to 0°C and n-BuLi (11 mmol) was added slowly over five min. The resulting solution was stirred for 30 minutes at 0°C prior to cooling to −78°C. Then, the ester (10 mL) was added slowly over 5 min. The resulting solution was stirred for an additional 30 minutes at −78°C. Then, a solution of trimethylsilyl chloride (11 mmol) and HMPA (22 mmol) in THF (4 mL) were added slowly via cannula. The dry ice/isopropyl alcohol bath was then removed and the solution was allowed to stir for 1 h at room temperature. The yellow solution was diluted with cold pentane (50 mL) and washed with cold water (3 × 75 mL). The organic layer was dried over sodium sulfate (10 g) and filtered through Celite and concentrated *in vacuo*. The residue was purified by Kügelrohr to give the silyl enol ether product in 42% yield and an (E/Z) ratio of 82/18. Then, a flame-dried, 20 mL, 2-neck flask containing a solution of the bis-phosphoramide (0.1 mmol) in dichloromethane (10 mL) was cooled to −78°C under nitrogen and benzaldehyde (2.0 mmol) was added in one portion. To the resulting solution, $SiCl_4$ (2.2 mmol) was added and the reaction mixture was allowed to stir at −78°C for 5 minutes. Then the silyl enol ether (2.4 mmol) was added to the reaction mixture. The resulting mixture was stirred at −78°C for 15 min and the cold reaction mixture was poured into a rapidly stirring solution of 1/1 saturated $KF/1.0$ M KH_2PO_4 solution (50 mL). The biphasic mixture was stirred vigorously for 2 h after which the aqueous layer was washed with dichloromethane (3 × 50 mL). The combined organic extracts were dried over sodium sulfate, filtered and the filtrate was concentrated *in vacuo*. The residue was purified by silica gel column chromatography to give the coupled products in 97% yield.

Alcohol-Ketones

See Ketone-Alcohol.

Aldehydes and Ketones

58 From Aldehydes or Ketones and Alkyl Halides

Alkyl halides react with enolate anions of aldehydes or ketones by an S_N2 reaction to give the alkylated product. Enolate anions of aldehydes and ketones react with alkyl halides to give alkylated aldehydes or ketones.

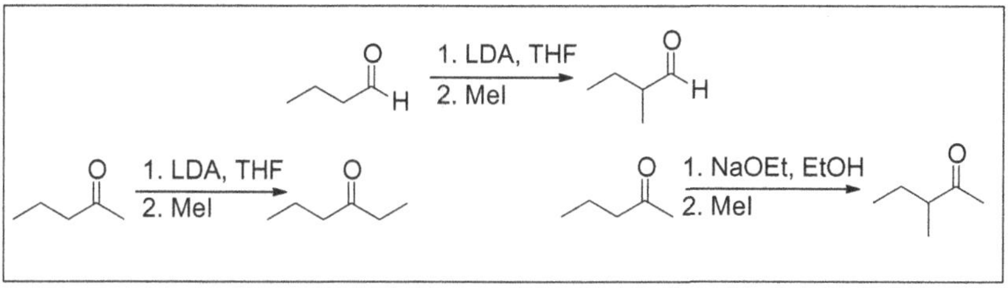

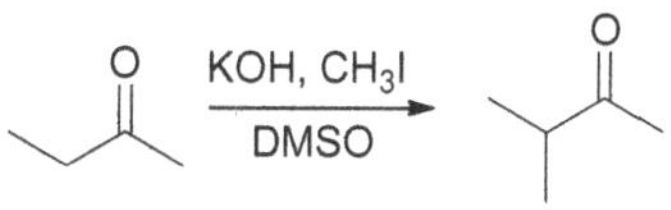

Tetrahedron Lett. **1990**, 31, 859–862.

In a three-necked flask containing a reflux condenser, a mechanical stirrer and a dropping funnel a mixture of butan-2-one (0.1 mol) and methyl iodide (0.8 mol) is added to a suspension of potassium hydroxide (2 mol) in dimethyl sulfoxide (190 mL) under stirring at 50−60°C. After initial heating the temperature of the reaction mixture is kept within this temperature range by means of the exothermic reaction. After the addition, stirring is continued for an hour, and the slurry is poured into 500 mL of ice water. The product is extracted with pentane (3 × 70 mL), the organic layer washed twice with water and dried over magnesium sulfate. Evaporation of the solvent gives 3-methylbutan-2-one, which can be further purified by distillation.

59 From Organometallics and Dimethylformamide (DMF)

Aryl organometallics react with DMF via acyl substitution to give aldehydes.

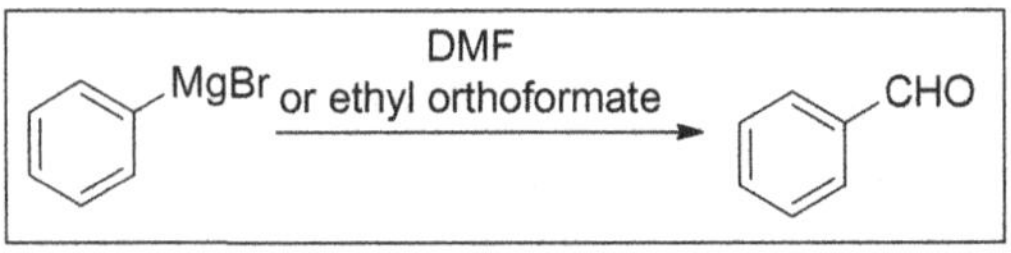

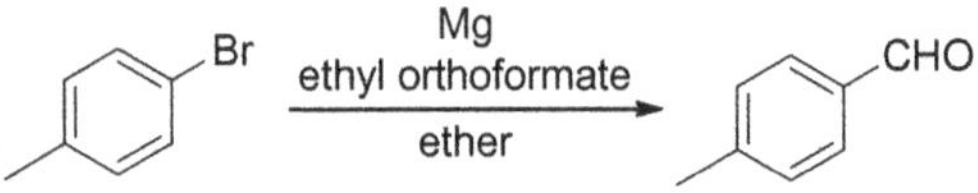

J. Org. Chem. *1941*, 6, 489–506.

The Grignard reagent was prepared in the usual way under nitrogen by adding a solution of p-bromotoluene (0.122 mol) in ether (100 mL) dropwise over 45 min and with stirring to magnesium powder (0.122 mol) suspended in ether (20 mL). The reaction was started by addition of ethyl bromide (0.5 mL) and a crystal of iodine and after all of the halide was added, the solution was refluxed for two h. Ethyl orthoformate (0.142 mol) in ether (30 mL) was then added over 5 min and the reaction mixture was refluxed for 5 h. The ether was then distilled off on the steam bath and when practically all of it was removed, there was a sudden vigorous reaction. At this point, the flask was quickly immersed in an ice-bath and allowed to remain there until all evidence of a reaction had disappeared. After standing overnight, ice (50 g) and cold 5 N hydrochloric acid (125 mL) were added, the small residual amount of ether was evaporated and the reaction mixture was refluxed for thirty min on the steam bath under an atmosphere of carbon dioxide. The aldehyde was then steam distilled in an atmosphere of carbon dioxide and the distillate was extracted with ether (3 × 60 mL). The combined ether extracts were evaporated on the steam bath to remove solvent and propionic aldehyde (b.p. 50°C) and the residue of p-tolualdehyde was taken up in ether (20 mL). The ethereal solution was then vigorously shaken with a freshly prepared saturated aq solution of sodium bisulfate and the solid was filtered off, and the filtrate was shaken again with fresh bisulfite solution and filtered. The combined solids were washed with ether and dried. The substance weighed 20.3 g (74.4 %).

Aldehyde-Alkenes

60 From Allylic Vinyl Ethers

Allyl vinyl ethers undergo Claisen rearrangement to give alkenyl aldehydes or alkene ketones. Alkene-esters are converted to a silyl enol ether and Claisen rearrangement followed by hydrolysis leads to a new alkene acid.

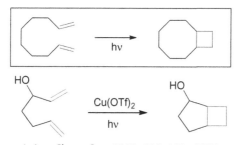

J. Am. Chem. Soc. *1999*, 121, *10865–10874*.

A mixture of allyl vinyl ether (0.895 mol), *p*-cymene (1.5 L), and 7.42 g of diglyme (internal standard) was heated to ~120°C under nitrogen with stirring. Aliquots were periodically removed, and GC was used to monitor the progress of the reaction. After 22 h, the conversion was ~87%. Vacuum transfer of the cooled reaction mixture followed by fractional distillation on a glass-bead column afforded 4.025 g of pent-4-enal in >99% purity (GC).

Alkanes

61 From Alkenes

[2 + 2]-Photocycloaddidtion generates cyclobutane derivatives.

J. Am. Chem. Soc. *1982*, 104, *998–1007*.

A representative procedure for photobicyclization: 3-Hydroxyhepta-1,6-diene (0.17 mol) in ether (200 mL) with $Cu(OTf)_2 \cdot C_6H_6$ (0.8 mmol, 0.9 mol% CuOTf) was irradiated for 21 h with an internal 450-W Hanovia mercury vapor lamp. The resulting solution was poured into a mixture of ice (100 g) and concentrated NH_4OH (100 mL) and dried (Na_2SO_4). Solvent was removed by rotary evaporation and the product isolated by distillation under reduced pressure.

62 From Alkyl Halides and Organometallics

Grignard reagents and organolithium reagents react only with very reactive alkyl halides to give coupling products; react with alkyl halides in the presence of copper or iron salts to give coupling products. Organocuprates react with a variety of alkyl halides, vinyl halides, and aryl halides to give the coupling product.

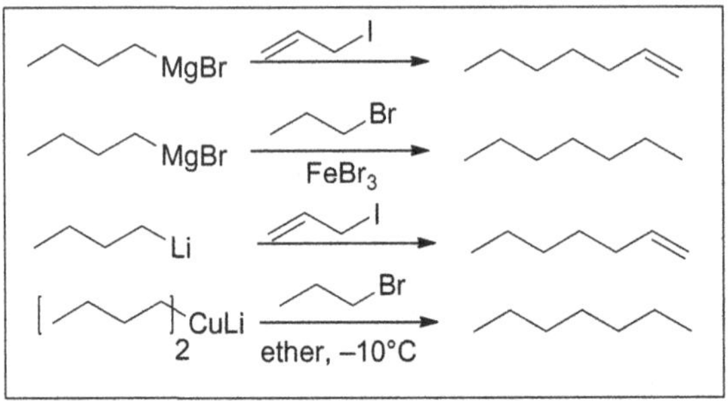

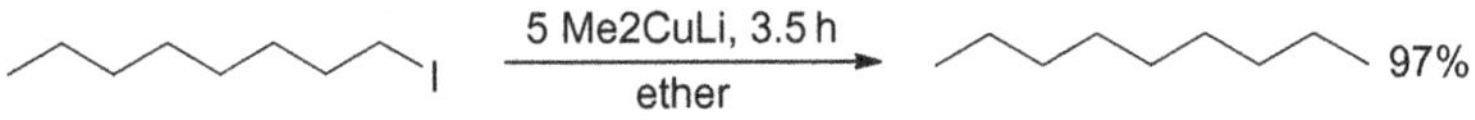

J. Am. Chem. Soc. **1969**, 91, 4871–4882.

Ether solutions (6.0 ml) containing 2.00 mmol of iodooctane, a weighed amount of n-decane (internal standard), and the various organometallic reagents specified below were stirred at room temperature and aliquots were periodically removed, hydrolyzed, and analyzed. In the reaction with 2.00 mmol of lithium dimethylcuprate, after 2.67 h the calculated yield of n-nonane was 97% and only 3% of the n-octyl iodide remained. In a similar reaction where 4.00 mmol of methyllithium was treated with only 0.10 mmol (5 mol%) of copper(I) iodide, the calculated yields after a 12-h reaction period were 19% n-octane, 6.5%, oct-1-ene, 64% n-nonane, and 2% of the starting iodide.

Alkenes

63 From Alkenes and Aryl Halides (Heck reaction)

Aryl halides are coupled to alkenes in the presence of palladium (0) catalyst.

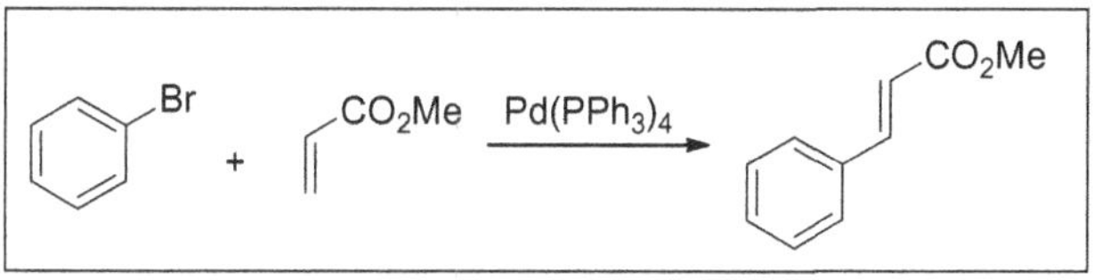

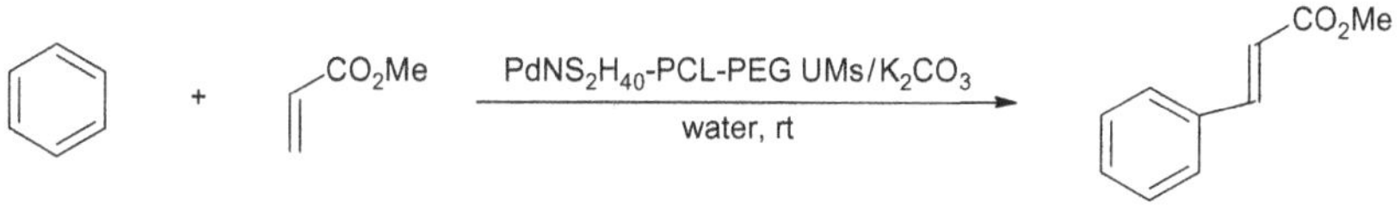

Applied catalysis A: General **2014**, 469, *183–190.*

At first, aryl halide (1 mmol) and methyl acrylate or styrene (1 mmol) were dissolved in water (2 mL). This reactant mixture was then added to the solution of K_2CO_3 (1.2 mmol) in water (1 mL) under stirring conditions. A Pd (0) (0.1 mol%) nanocatalyst (PdNs-H40-PCL-PEG Ums) was added to this mixture. The reaction was monitored by TLC, or by GC if necessary. On completion of the reaction, the reaction mixture was transferred into ultracentrifuge and the lower phase, which contains catalyst, was removed. Then, the mixture was extracted with dichloromethane. The organic phase was dried over sodium sulfate, filtered and evaporated in vacuum. The mixture was then purified by column chromatography over silica-gal or recrystallization to afford a product with high purity. Extraction with ethyl acetate or dialysis against water can also be applied to remove the catalyst.

64 From Alkenes and Dienes

Alkenes undergo Diels-Alder reactions with 1,3-dienes to give cyclohexenes.

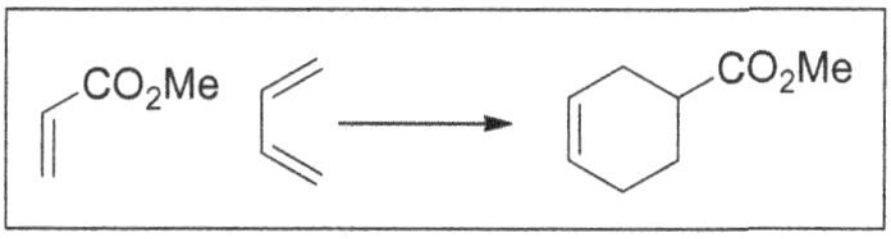

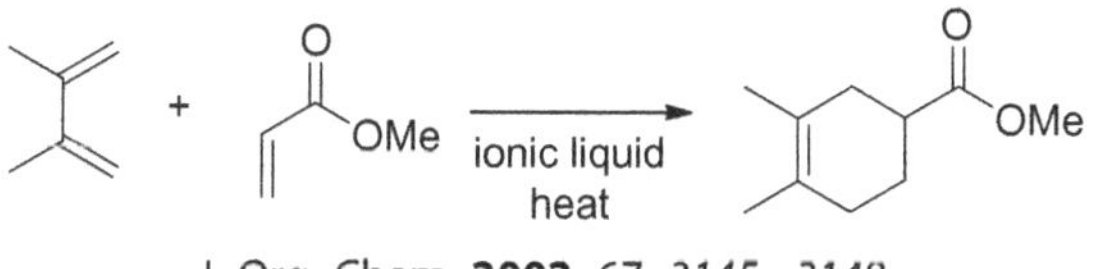

J. Org. Chem. **2002**, 67, *3145–3148.*

To a microwave vessel containing toluene (2 mL) and an ionic liquid (55 mg) was added 2,3-dimethylbuta-1,3-diene (2 mmol) and methyl acrylate (2 mmol). A Teflon coated magnetic stirrer bar was added and the vessel and sealed. The sample was irradiated for 5 minutes using a power of 100 W. For safety, the maximum pressure was set to 200 bar. After being cooled to room temperature, the vessel was opened and the toluene layer was removed using a pipet. The ionic liquid layer was washed with toluene and these washings were combined with the original toluene layer. Removal of the solvent resulted in a light yellow oil, the NMR spectrum of which showed product but no evidence for ionic liquid or its decomposition product but no evidence for ionic liquid or its decomposition products but gave 80% yield of product.

65 From Dienes

Cyclic, nonconjugated dienes are converted to cyclic dienes with Grubbs' I or II catalysts or the Shrock catalyst (Ring Closing Metathesis).

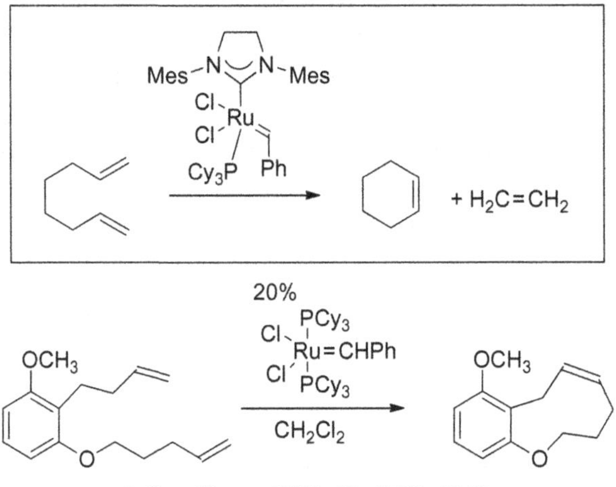

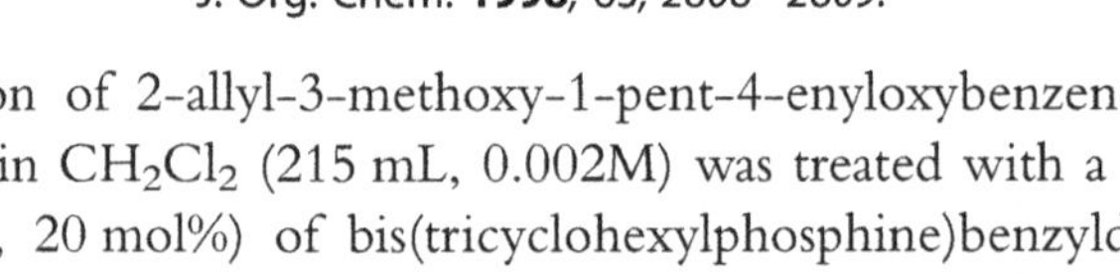

J. Org. Chem. **1998**, 63, 2808–2809.

A solution of 2-allyl-3-methoxy-1-pent-4-enyloxybenzene (100 mg, 0.43 mmol) in CH_2Cl_2 (215 mL, 0.002M) was treated with a single portion (71 mg, 20 mol%) of bis(tricyclohexylphosphine)benzyldine ruthenium (IV)dichloride. This solution was then gently refluxed for 16 h. The solvent was removed *in vacuo* and the brown residue slowly filtered through a short pad of Florisil using 10% EtOAc in hexane as the eluant. The eluate ($\approx$60 mL) was then concentrated and subjected to preparative layer chromatography to yield 8-methoxy-3,4,7-trihydro-2H-benzo[b] oxonine (40 mg, 45%) as a colorless oil.

66 From Ketones or Aldehydes and Ylids

Aldehydes and ketones react with phosphonium ylids to give alkenes by the Wittig reaction.

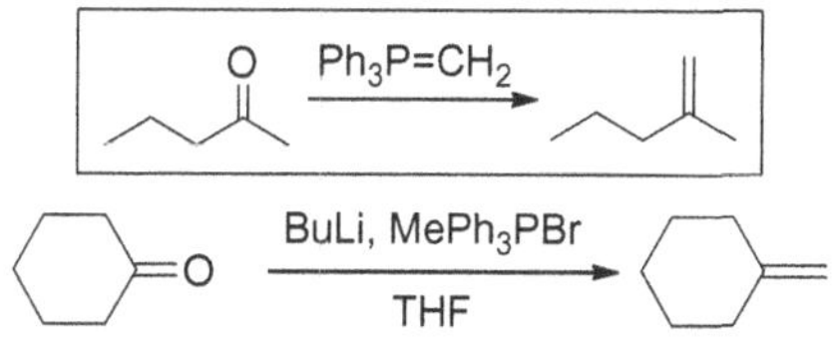

Org. Synth. **1960**, 40, 66.

A 500 mL 3-necked round-bottom flask was fitted with a reflux condenser, an addition funnel, a mechanical stirrer and a gas inlet tube. A gentle flow of nitrogen through the apparatus was maintained throughout the reaction. An ethereal solution or *n*-butyllithium (0.10 mol) and anhydrous ether (20 mL) was added to the flask. The solution was stirred and triphenylmethyl phosphonium bromide (35.7 g) was added cautiously over 45 min. The solution was stirred at room temperature for 4 h. Freshly distilled cyclohexanone (0.11 mole) was added dropwise. The solution became colorless and a white precipitate separated. The mixture was heated at reflux overnight, cooled to room temperature and the precipitate was removed by suction filtration. The precipitate was washed with 100 mL of ether and the combined ethereal filtrates extracted with 100 mL portions of water until neutral, and then dried over calcium chloride. The ether was carefully distilled through an 80 cm column packed with glass helices. Fractionation of the residue remaining after removal of ether through an efficient at low holdup column gave 3.4–3.8 g of pure methylene cyclohexane.

Alkynes

67 From Alkyne Anions and Alkyl Halides (S_N2 Reaction) or Aryl Halides

Terminal alkynes react with a strong base to give an alkyne anion, which reacts with an alkyl halide via an S_N2 reaction, to give an internal alkyne. Aryl halides are coupled to alkynes in the presence of a palladium (0) catalyst to give alkynes **(Sonogashira Coupling)**.

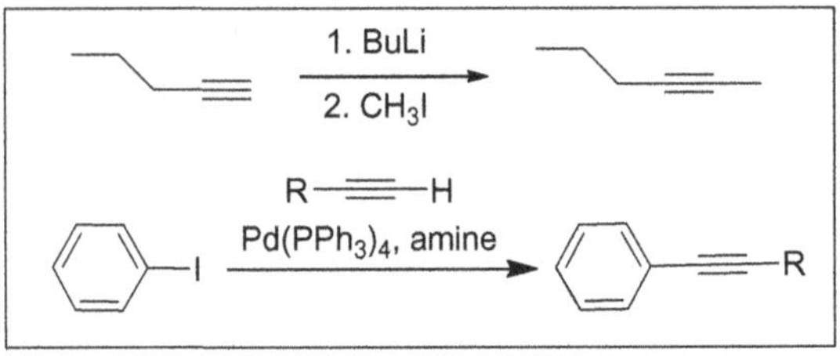

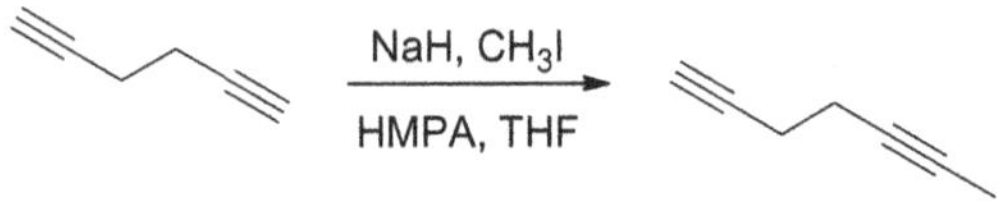

J. Org. Chem. **1980**, *45, 2259–2261.*

To a magnetically stirred oil-free suspension of sodium hydride (0.115 mol) in dry THF (200 mL) were added hexa-1,5-diyne (0.064 mol), of HMPA (0.014 mol) and dry methyl iodide (0.257 mol), *via* syringe. After 72 h, the reaction was quenched by addition of water (100 mL). The products were isolated by extraction into pentane followed by fractional distillation through a 10 cm Vigreux column, affording 3.45 g (59%) of hepta-1,5-diyne as a colorless liquid.

J. Am. Chem. Soc. **2013**, *135, 10829–10836.*

To a solution of 2-methyl-2-(prop-2-yn-1-yl) cyclopentane-1,3-dione (10.0 mmol), $Pd(OAc)_2$ (0.25 mmol), PPh_3 (1.0 mmol), CuI (0.5 mmol), and triethylamine (17.2 mmol) in DMSO (20 mL) at room temperature was added a solution of 1-bromo-4-nitrobenzene (11 mmol) in DMSO (10 mL) via cannula and the mixture was stirred at 90°C for 2 h. The reaction was cooled to room temperature, diluted with water (50 mL) and extracted with ether (50 mL). The organic phase was washed with 10% HCl aq solution (3 × 20 mL), dried with sodium sulfate, filtered and concentrated *in vacuo*. The residue was purified by column chromatography.

Alkyne-Alcohols

68 From Ketones or Aldehydes and Alkynes

Aldehydes and ketones react with alkyne anions to form a new C-C bond and form an alkyne-alcohol.

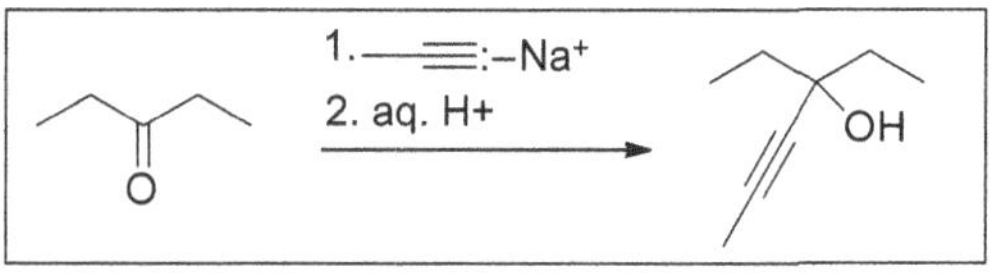

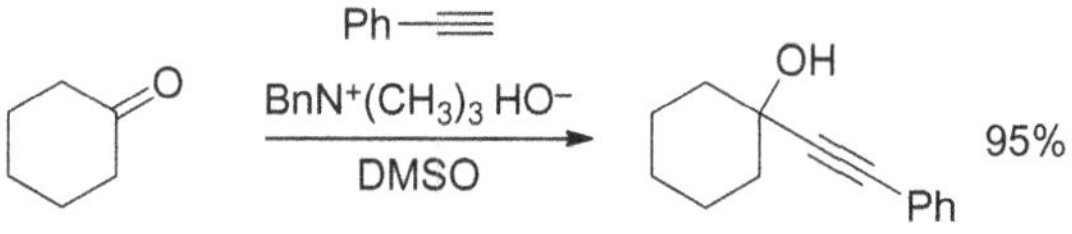

J. Org. Chem. **2003**, 68, 3702−3705.

To a solution of cyclohexanone (5.0 mmol) and phenylacetylene (6.0 mmol) in DMSO (2.5 mL) was added a solution of 3 (0.5 mmol) in DMSO (2.5 mL) over 10 min at room temperature. The reaction was stirred until the disappearance of starting materials as indicated by TLC. To the reaction mixture was added water (20 mL), and the mixture was extracted with ethyl acetate-hexane (1:3). The combined organic layers were dried over anhydrous Na_2SO_4 and concentrated to give a crude residue, which was purified by column chromatography on silica gel using ethyl acetate-hexane to afford 1-(phenylethynyl)cyclohexan-1-ol.

Arenes

69 From Aromatics and Alkyl Halides

Alkyl halides react with Lewis acids to give carbocations, which react with benzene to give arenes (Friedel-Crafts Alkylation).

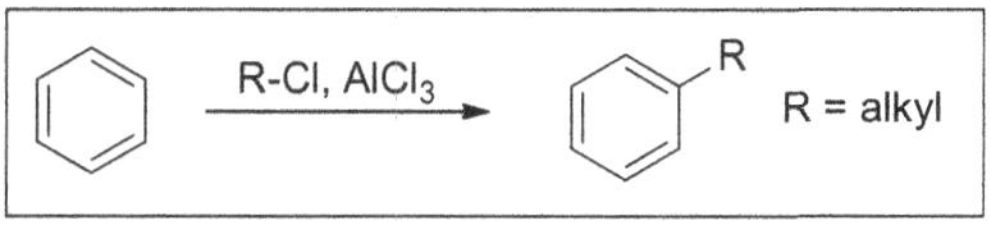

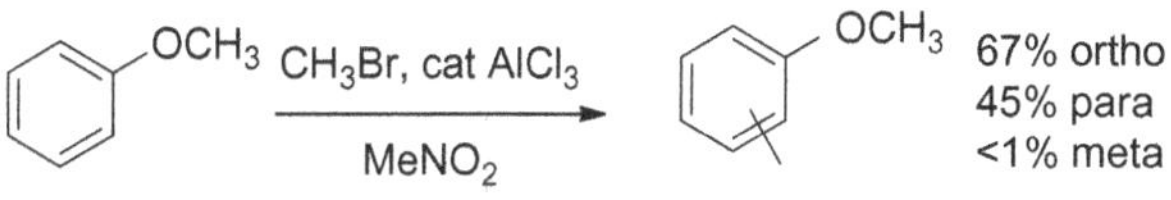

J. Am. Chem. Soc. **1984**, 106, 5284−5290.

Anisole (10.8 g, 0.1 mol) and 2.7 g of AlCI$_3$ or 1.4 g of BF$_3$ (0.02 mol) were dissolved in 20 mL of nitromethane. While the solution was kept at 25°C with good stirring, 0.0902 mol of bromomethane dissolved in 10 mL of nitromethane was added. The reaction mixture was reacted for 30 min at 25°C. It was thereafter quenched with ice water and the organic layer separated, extracted with ether, washed, dried and analyzed by gas liquid chromatograph.

Biaryls
70 From Aryl Halides (Suzuki-Miyaura Coupling)
Aryl triflates or halides couple to aryl boronic acids in the presence of a palladium(0) catalyst to give biaryls.

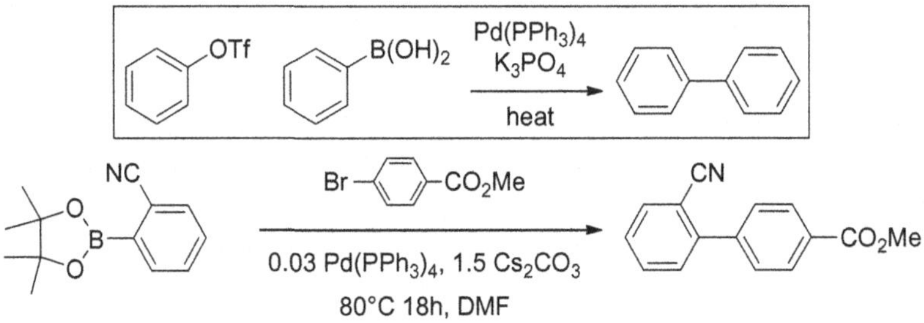

J. Org. Chem. **2003**, 68, 3729–3732.

To a 50-mL three-necked round-bottom flask, were charged, in no specific order (4′-bromoacetophenone (2.6 mmol), bis-(pinacolato) diboron (2.8 mmol), palladium acetate (0.078 mmol), potassium acetate (7.9 mol) and 10 mL of DMF. The mixture was degassed by gently bubbling argon through for 30 minutes at room temperature. The mixture was then heated at 80°C under argon until completion of the reaction (2–3 h). After the reaction mixture was cooled to room temperature, 1-bromo-4-nitrobenzene (2.6 mmol), cesium carbonate (3.9 mmol) and Pd(PPh$_3$)$_4$ (0.078 mmol) were added. The reaction mixture was then heated at 80°C overnight under argon, then cooled to room temperature and diluted with water (20 mL) and ethyl acetate (20 mL). Black particles were removed by passing through a pad of Celite. The organic layer was separated, and washed twice with 15 mL of brine solution. After drying over sodium sulfate, the solvent was removed at reduced pressure to afford the biphenyl product in >99% yield.

Carboxylic Acids

71 From Carbon Dioxide and Organometallics

Alkyl halides react with Mg or Li and then CO_2 to give a carboxylic acid.

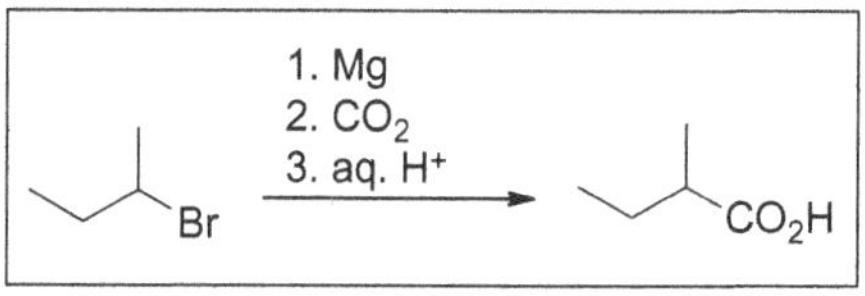

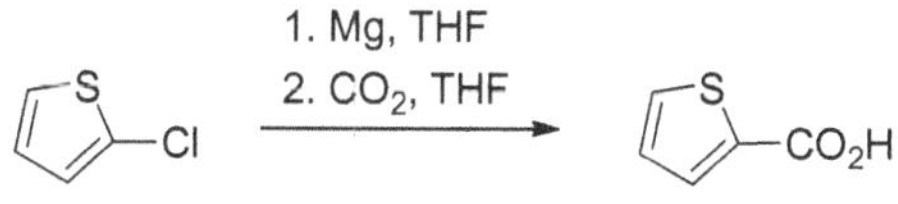

Org. Biomol. Chem. **2010**, 8, 1688–1694.

Magnesium turnings (18 mmol) and 2-chlorothiophene (14 mmol), dissolved in dry THF (10 mL) were introduced in an oven-dried 25 mL three neck round-bottomed flask under an argon atmosphere. The mixture was heated by a preheated oil bath (85°C) to reflux. Additional heating was applied for 30–200 min. The solution of Grignard reagent was separated from the remaining Mg with a syringe.

The solution of the Grignard reagent was added dropwise to an excess of freshly prepared dry, solid CO_2. After addition of the reagent another layer of dry CO_2 was deposited. The mixture was slowly heated to room temperature in 15 min, acidified with 10% (by wt) HCl (10 mL) and extracted with toluene (3×10 mL). The organic layer was dried with magnesium sulfate, filtered and evaporated.

Dienes

72 From Dienes

1,3-Dienes can undergo 1,5-sigmatropic hydrogen shifts to give new 1,3-dienes. 1,5-Dienes undergo Cope rearrangement to give new 1,5-dienes.

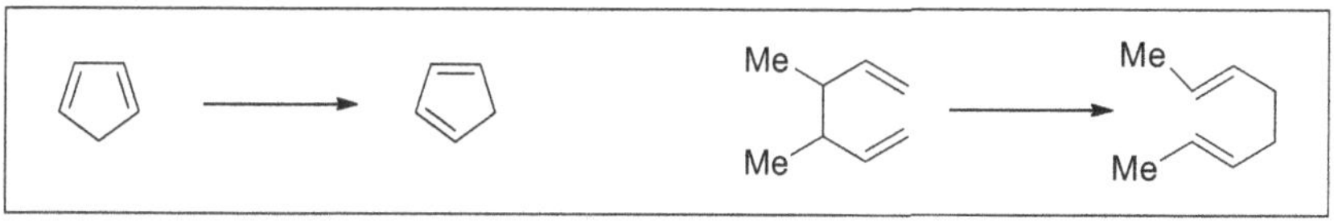

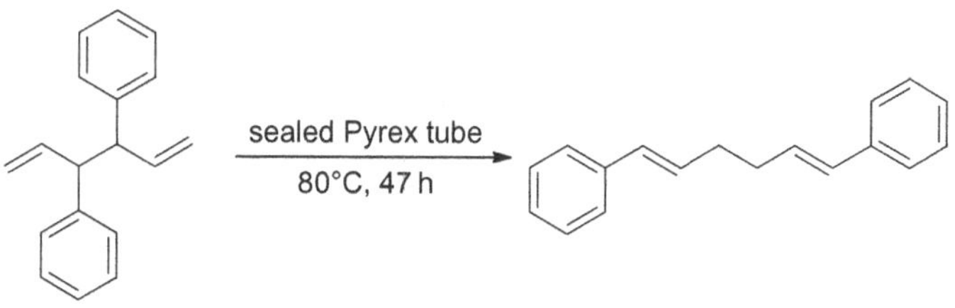

J. Am. Chem. Soc. **1971**, 93, 3985–3990.

A 106-mg sample of 3,4-diphenylhexa-1,5-diene was sealed into an evacuated ($\approx$0.1 mm) Pyrex tube and maintained at 80°C for 47 h. A reaction time of 47 h corresponds to about six half-lives, which means that 1.5% of the diene remained unreacted. The composition of the crude product, by glpc, was 99.1% trans, trans-1,6-diphenylhexa-1,5-diene and 0.9% of a component thought to be an impurity in the starting material. The cis, cis isomer as well as trans-1,4-diphenylhexa-1,5-diene could not be detected under conditions where about 0.1% would be observed. Recrystallization of the crude product from methanol gave material having mp 79.0−79.5°C which, on mixing with authentic (mp 79.0−79.5°C) trans, trans-diene, gave an undepressed melting point, 79.2−79.8°C. An infrared spectrum of the recrystallized material was virtually identical with that of authentic trans, trans-diene.

Diketones

73 From Ketones and Esters

Enolate anions of aldehydes and ketones react with esters to give β-diketones.

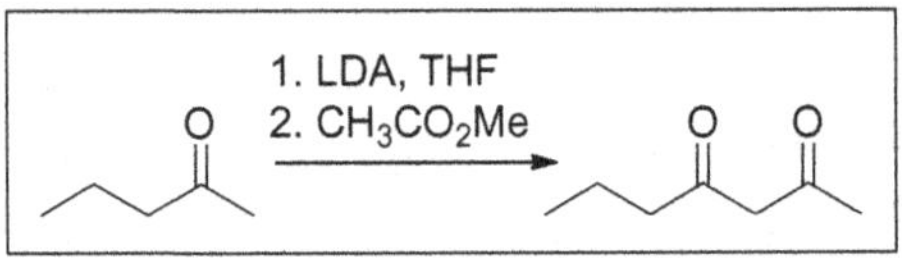

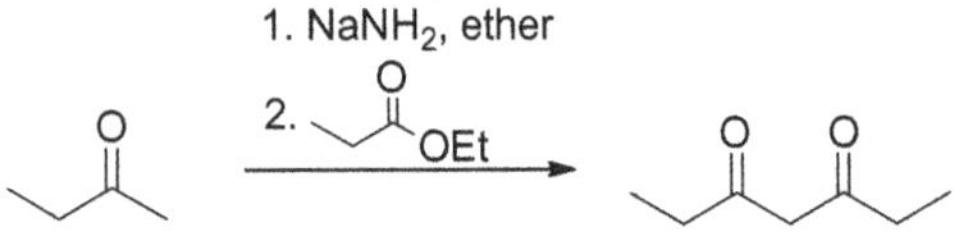

J. Am. Chem. Soc. **1944**, 66, 1220−1222.

To the stirred suspension of sodium amide was added, during five to ten min, a solution of the calculated amount of butan-2-one in 50 mL of absolute ether. After five min (when the sodium derivative of the ketone was assumed to be formed) the calculated amount of ethyl propanoate in 50 mL of absolute ether was added, and the stirring continued for 2 h while the mixture was heated at reflux on a steam-bath. The mixture, containing a gelatinous precipitate of the sodium salt of the β-diketone, was poured into 300 mL of water, neutralized with dilute hydrochloric acid and extracted with ether. The solvent was distilled from the ether solution and the residue dissolved in an equal volume of methanol. The β-diketone was precipitated from this methanol solution in the form of its copper salt, from which the free β-diketone was regenerated by treatment with acid (61%) yield.

Esters

74 From Esters and Alkyl Halides

Alkyl halides react with enolate anions of esters by an S_N2 reaction to give the alkylated product.

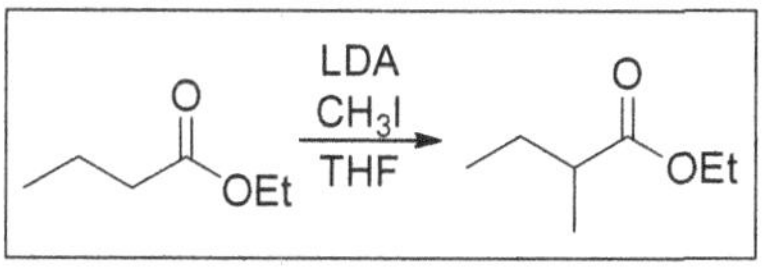

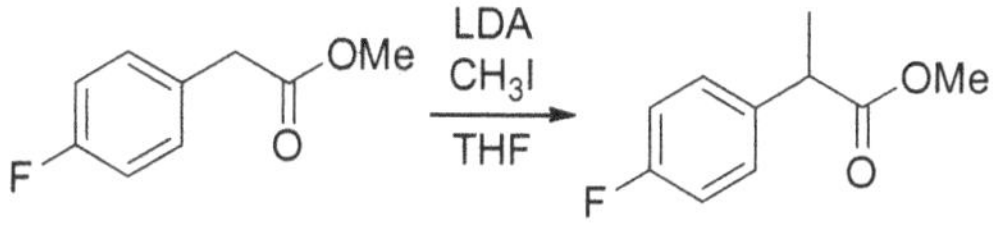

J. Org Chem. **2003**, 68, 7234–7242.

To a solution of diisopropylamine (3.6 mmol) in THF (4 mL) was added n-butyllithium (3.65 mmol) at −78°C under Ar and the mixture was stirred for 30 min. Then a solution of methyl 4-fluorophenylacetate (3.0 mmol) in THF (5 mL) was added dropwise, and the mixture was stirred for 20 min. at the same temperature. The mixture was then warmed to 0°C. Methyl iodide (4.8 mmol) was added and the mixture was stirred for 30 min. The reaction mixture was acidified by 2 M HCl and extracted with ethyl acetate. The organic layer was washed with brine, dried over anhydrous sodium sulfate and concentrated *in vacuo*. The residue was purified by silica gel column chromatography (hexane/ethyl acetate 19/1) to give methyl 2-(4-fluorophenyl) propanoate (0.48 g, 89% yield) as a colorless oil.

75 From Esters

Conjugated esters undergo primarily 1,4-addition with malonate anion derivatives. (conjugate addition; Michael addition). Esters react under kinetic and thermodynamic conditions to give β-keto esters via self–condensation (Claisen condensation). α,ω-Diesters react with base to give cyclic ketone-esters (an intramolecular Claisen condensation, which is called the Dieckmann condensation).

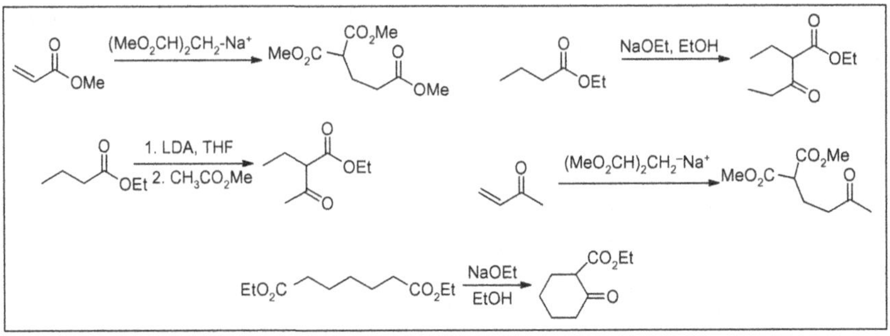

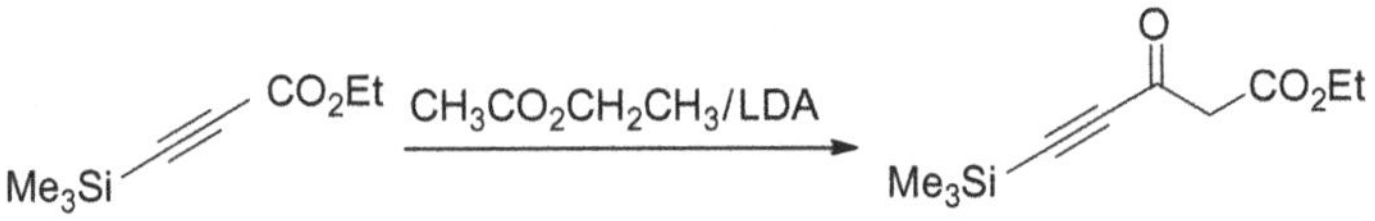

To a cooled (−78°C) solution of LDA (2.0 M in heptane/THF/ethylbenzene, 2.74 mL, 5.47 mmol, 5.0 eq.) in THF (8.0 mL) was added dropwise dry *tert*-butyl acetate (753 mL, 5.58 mmol, 5.1 eq.). After stirring at that temperature for 1 h, a solution of ethyl 3-(trimethylsilyl)propiolate (250 mg, 1.10 mmol, 1.0 eq.) in THF (1.0 mL and 1.0 mL wash) was added dropwise. The mixture was stirred at −78°C for 1 h, quenched by addition of 1.0 M aq. NaH_2PO_4 (3.0 mL). While warmed to room temperature, brine (10.0 mL) and diethyl ether (10.0 mL) were added. The layers were separated and the aq layer was extracted with diethyl ether (2 × 20.0 mL). The combined organic phases were dried over $MgSO_4$, and concentrated *in vacuo*. Ethyl 3-oxo-5-(trimethylsilyl)pent-4-ynoate was isolated by column chromatography (silica gel, 3.0 g, hexanes/EtOAc = 40:1 to 20:1) as a light yellow liquid, as 10:1 mixture of keto- and enol- isomers, 273 mg, 0.915 mmol, 83%.

Ketones

76 From Acid Derivatives

Acid chlorides, acid anhydrides react with organocuprates, and Weinreb amides react with Grignard reagents to give ketones.

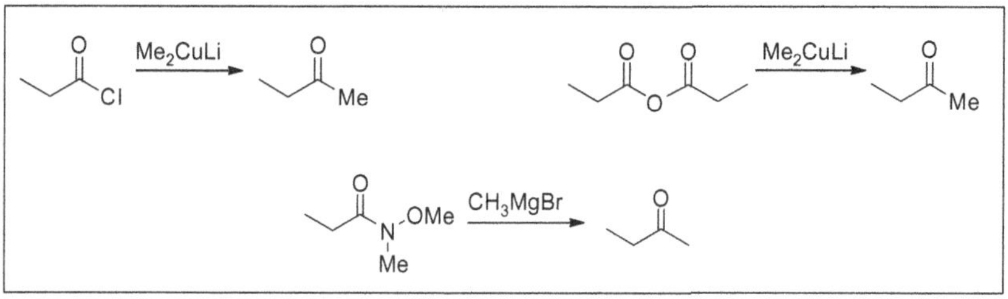

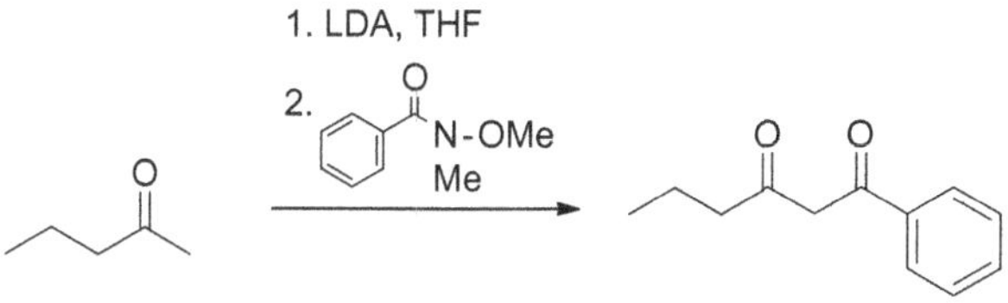

J. Org. Chem. **1989**, 54, 4229–4231.

A solution of diisopropylamine (11−15 mmol) in 20 mL of THF or ether was cooled under nitrogen to −78°C and n-butyllithium (11−15 mmol) in hexane was added. The appropriate ester (11−15 mmol) in 1 mL of THF or ether was added dropwise and after 15 minutes at −78°C, a solution of the N-methoxy-N-methylamide (10 mL) in 1 mL of THF or ether was slowly added. The reaction mixture was stirred at −78°C for the specified period of time, warmed to near room temperature and poured into 1 N HCl. The mixture was extracted twice with ether, and the combined organic layers were dried over MgSO₄ and evaporated to dryness. The residue was chromatographed through a short column of silica gel, eluting with 9:1 hexane: ethyl acetate and then distilled in a Kügelrohr apparatus.

77 From Aromatics and Acid Derivatives
Benzene reacts with acid chlorides in the presence of a Lewis acid catalyst to give ketones via electrophilic aromatic substitution (Friedel-Crafts Acylation).

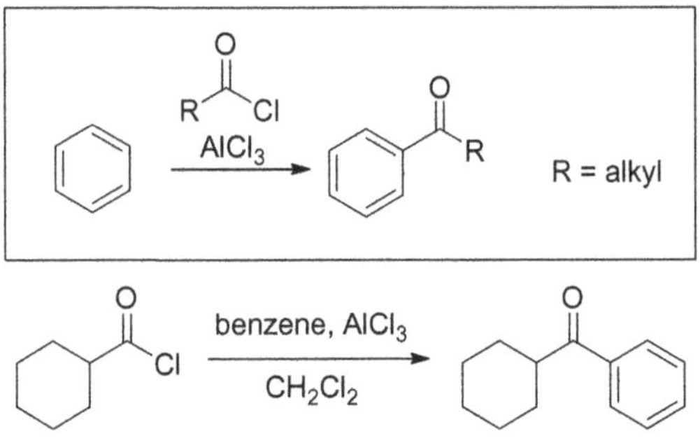

PATENT: WO 01/85698.

A solution of cyclohexanecarbonyl chloride (9.1 mL) in dichloromethane (25 mL) was added slowly under nitrogen atmosphere at 0−4°C to a stirred mixture of AlCl₃ (9.1 g), dichloromethane (25 mL) and benzene (50 mL). The resulting mixture was stirred for 1 h at 0−4°C and 12 h at room temperature. The mixture was poured into ice-water (200 mL, contains 1 mL of concentrated HCl) and stirred for 5 min. The phases were separated and the aqueous phase was washed with dichloromethane (2 × 20 mL). The organic phases were combined and extracted with water (2 × 20 mL), 2.5% NaOH solution (2 × 30 mL) and water (2 × 20 mL). The organic phase was dried over sodium sulfate and evaporated. Yield was 12.0 g. Note that benzene is a cancer suspect agent.

78 From Ketones and Conjugated Ketones with Organometallics

Conjugated ketones and aldehydes undergo primarily 1,4-addition with organocuprates. Conjugated ketones undergo a mixture of 1,2- and 1,4-addition with Grignard reagents. Conjugated ketones and aldehydes react with ketones under thermodynamic enolate anion conditions to give cyclic ketones, and with cyclic ketones to give bicyclic ketones (Robinson Annulation).

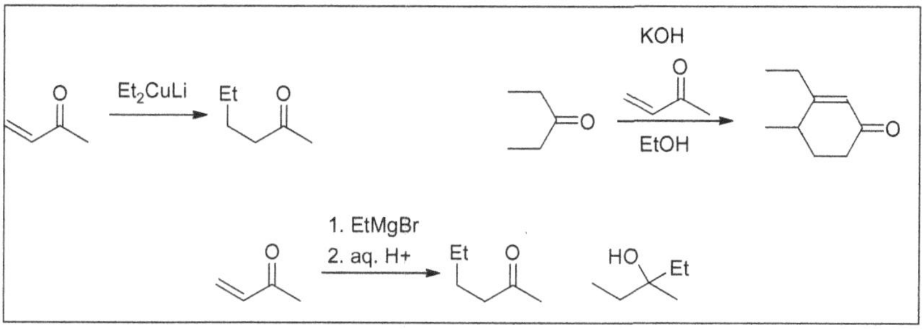

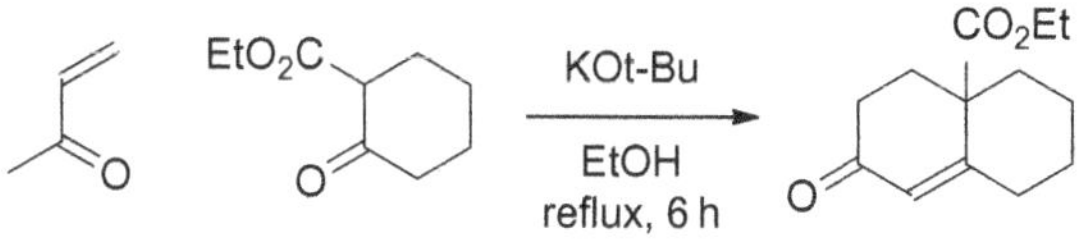

Org. Lett. **2003**, 5, 137–140.

Potassium *tert*-butoxide (296 mg, 2.65 mmol) was dissolved in ethanol (EtOH, 25 mL) at 0°C, under argon. After stirring for 20 min, ethyl 2-oxocyclohexanecarboxylate (8 mL, 50 mmol) was added slowly at the same temperature. After 15 min at 0°C, methyl vinyl ketone (4.15 mL, 50 mmol) was added over 5 h via syringe pump. Then the resulting deep-orange solution was heated to reflux and kept for 6 h. The reaction mixture was cooled to ambient temperature. After stirring for 18 h, the mixture was poured into a separatory funnel containing saturated NH_4Cl (30 mL) and extracted with Et_2O (3 × 200 mL). The combined organic layers were dried over $MgSO_4$ and solvent removal afforded ethyl 7-oxo-1,3,4,5,6,7-hexahydronaphthalene-4a(2*H*)-carboxylate (11.0 g) as a crude orange oil, which was used without further purification.

79 From Nitriles and Organometallics

Nitriles react with Grignard reagents or organolithium reagents to give an imine, which is hydrolyzed to a ketone.

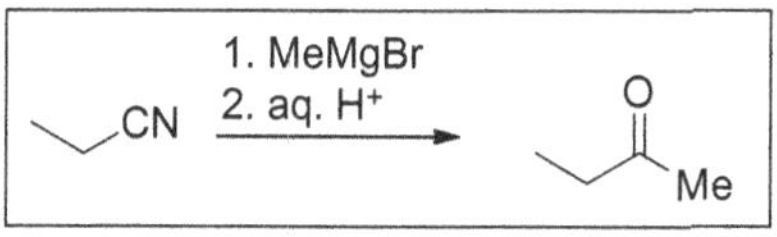

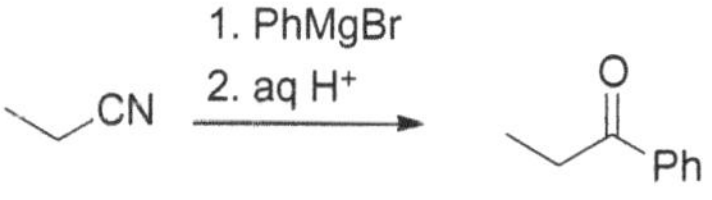

J. Am. Chem. Soc. **1930**, 52, 1267–1269.

The phenylmagnesium bromide was prepared in the usual way from 25 g of magnesium turnings and 160 g bromobenzene in 350 mL of dry ether. A solution of 0.25 mol of propanenitrile in 100 mL of dry ether was run in slowly with stirring during a period of fifteen min. The solution was stirred for an hour longer and allowed to stand overnight. The mixture was poured onto 500 g of ice and 30 mL of concentrated hydrochloric acid. The water layer, which contained the hydrochloride of the ketimide was evaporated from the ether layer and refluxed vigorously for 1 h. The solution is cooled and extracted with four 200 mL portions of ether. The ether extract was dried over anhydrous calcium chloride and the ether distilled from a water bath. The residue was transferred to a small modified Claisen flask and vacuum distilled.

Ketone-Alcohols

80 From Ketones and Aldehydes

Enolate anions of aldehydes and ketones react with aldehydes or ketones under kinetic or thermodynamic conditions to give β-hydroxy aldehydes or ketones by the aldol condensation at the less substituted α-carbon. (aldol condensation). An intramolecular aldol condensation leads to a cyclic β-hydroxy aldehydes or ketones. Silyl enol ethers react with aldehydes or ketone in the presence of a Lewis acid to give the aldol product (Mukaiyama aldol).

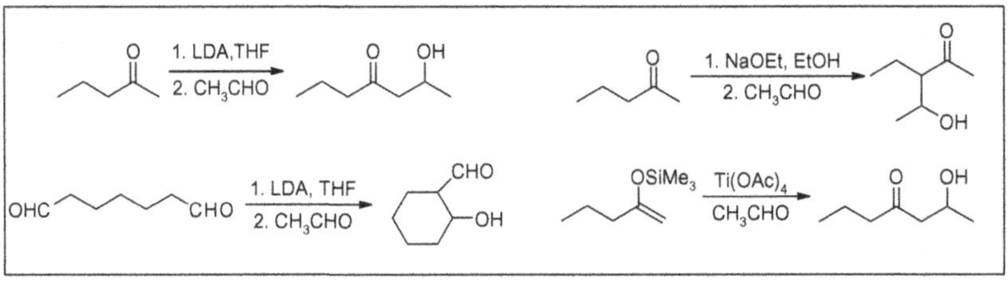

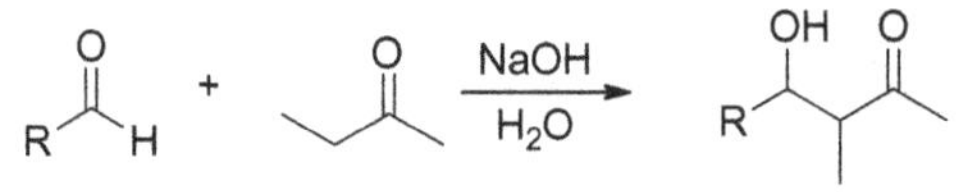

Org. Proc. Res. Develop. **2002** 6, 628–631.

An n-alkanal (0.1 mol) was added dropwise to a mixture of 2% aqueous sodium hydroxide (0.4 mol) and methyl ethyl ketone (0.4 mol) at 5°C or 20–25°C for 5–8 h. The reaction mixture was stirred at the same temperature for 15 h and then the organic layer was separated and washed with 5% NaCl solution. The solvent was removed under vacuum and the residue was distilled under reduced pressure to give a *syn/anti* mixture of 4-hydroxy-3-methyl-2-alkanone. The purities were 80–93% by GC.

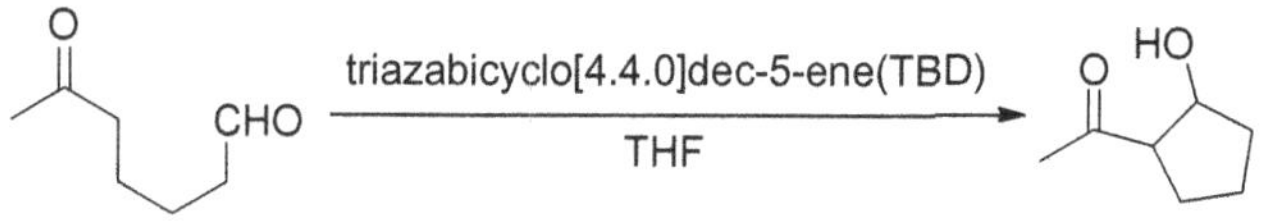

Eur. J. Org. Chem. **2008**, 4104–4108.

To a solution of ketoaldehyde (1.95 mmol) in anhydrous THF (6.5 mL) was added TBD (8 mol%) at room temperature. The reaction mixture was stirred for 30 min and then quenched with saturated ammonium chloride solution (10 mL). The organic layer was separated. The aqueous layer was extracted with diethyl ether (3 × 10 mL). The organic layer was dried with magnesium sulfate, filtered and concentrated *in vacuo* to afford a crude product that was purified by flash chromatography to give desired aldol products as mixtures of two diastereomers.

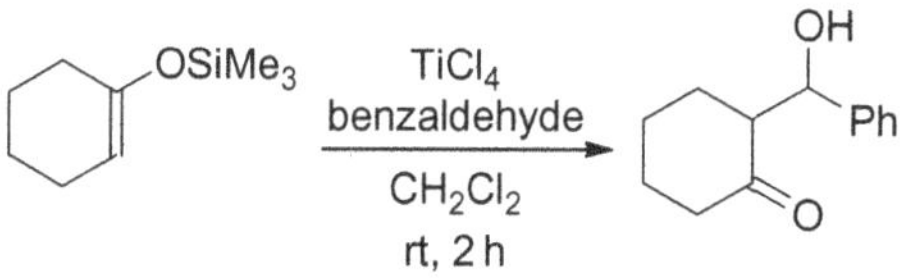

J. Am. Chem. Soc. **1974**, 96, 7503–7509.

A methylene chloride (10 ml) solution of 0.426 g (2.5 mmol) of 1-trimethylsilyloxycyclohex-1-ene was added dropwise into a mixture of 0.292 g (2.75 mmol) of benzaldehyde and 0.55 g (2.75 mmol) of $TiCl_4$ in dry methylene chloride (20 ml) under an argon atmosphere at −78°C, and the reaction mixture was stirred for 1 h. After hydrolysis at that temperature, the resulting organic layer was extracted with ether, and the extract was washed with water and dried over anhydrous Na_2SO_4. The mixture was condensed under reduced pressure, and the residue was purified by column chromatography (silica gel). Elution with methylene chloride afforded 115 mg (23%) of erythro-2-(1'-hydroxybenzyl)cyclohexan-1-one. From the last fraction, 346 mg (69%) of threo-2-(1'-hydroxybenzyl)-1-cyclohexanone was obtained.

KETONE-ESTER

Nitriles

81 From Alkyl Halides

Metal cyanides react with primary and secondary alkyl halides via a S_N2 reaction to give the corresponding nitrile. Alkyl halides undergo substitution reactions with NaCN or KCN (S_N2 reactions).

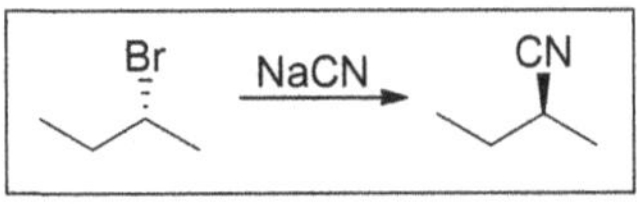

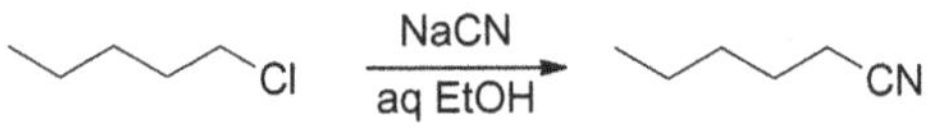

Ind. Eng. Chem. **1931**, 23, 352–353.

Weighed amounts of sodium cyanide, water, alcohol, and amyl chloride were placed in the order named in a 500-mL Pyrex Florence flask and fitted with a reflux condenser. The reaction was heated to reflux in a water bath, cooled, and the liquid contents of the flask were distilled out. The distillate was treated with anhydrous potassium carbonate, and the upper layer was separated, 95% ethanol was added to bring the total volume of ethanol up to 75 mL, and distilled. When no more distillate came over, the residue was cooled, treated with water in a separatory funnel to remove traces of alcohol, dried with calcium chloride, and weighed as crude amyl cyanide (60.5%).

Special note! The use of a cyanide salt is very dangerous and should be done with proper ventilation at all times and should never, even as waste, be added to mixtures that contain an acid since this results in the release of deadly HCN gas.

Nitrile-Alcohols
82 From Ketones or Alcohols
Aldehydes and ketones react with NaCN or KCN or HCN/cat. H$^+$ to form a new C—C bond and form a nitrile-alcohol.

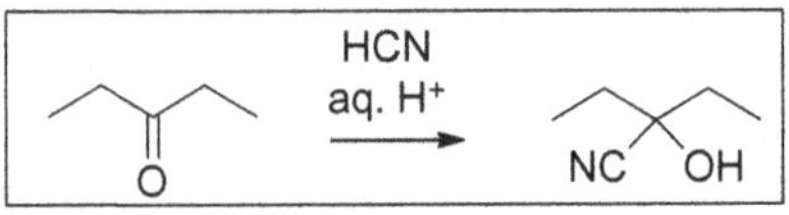

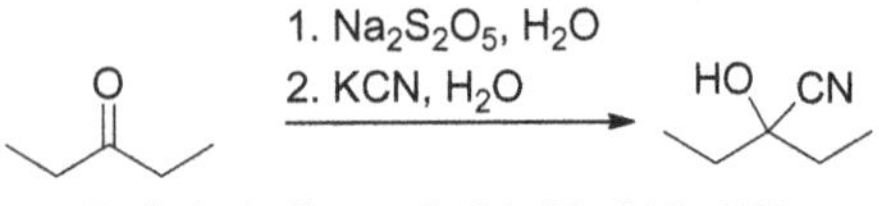

Turkish J. Chem. **2004**, 28, 345–350.

Sodium metabisulfite (11 g) was dissolved into 20 mL of cold water and placed into a 100 ml round-bottom flask. then, ketone (0.1 mol) was added slowly while swirling the liquid mixture steadily. This was followed by the addition of potassium cyanide (6 g) dissolved in 20 mL of cold water. During this slow addition, the cyanohydrin derivatives were separated out as the upper layer. When the separation was completed, the contents of the flask were transferred into a separating funnel and the lower layer was removed. The upper layer was transferred to a flask and sodium sulfate was then added to dry the product.

Special note! The use of a cyanide salt is very dangerous and should be done with proper ventilation at all times and should never, even as waste, be added to mixtures that contain an acid since this results in the release of deadly HCN gas. It is essential that the reader consult an MSDS or SDS prior to commencing any use of these chemicals.

INDEX

Note: Page numbers followed by "*f*" refer to figures.

Printed in the United States
By Bookmasters